American Institute of Chemical Engineers

Topical Conference on Innovative Environmental Management Procedures 2006

AA003471

Held at the 2006 AIChE Spring National Meeting

April 23-27, 2006
Orlando, Florida, USA

Printed from e-media with permission by:

Curran Associates, Inc.
57 Morehouse Lane
Red Hook, NY 12571
www.proceedings.com

ISBN: 978-1-60423-531-9

Some format issues inherent in the e-media version may also appear in this print version.

American Institute of Chemical Engineers

Topical Conference on Innovative Environmental Management Procedures
2006

Session 25 - Innovative Environmental Management Procedures Part 1

Cochairs: Javier Santillan
Jeffery Perl

Session 59 - Innovative Environmental Management Procedures Part II

Cochairs: Javier Santillan
Jeffery Perl

Session 137 - Innovative Environmental Management Procedures - Posters

Cochairs: Javier Santillan
Jeffery Perl

92 Concurrent Sustainable Environmental Program Management
Sekar Balusamy

93 Removal and Recovery of Heavy Metals from Industrial Waste Streams by Means of a Hybrid-Precipitation and Polymer Enhanced Ultrafiltration
Sezin Islamoglu, levent yilmaz

96 New Approach for Addressing Environmental Effects through Multi Objective Multi Refinery Optimization
Ahmed S. Khogeer, M. Nazmul Karim

97 Effective Field Characterization of a Groundwater Tce in Fractured Basalt
Sheri L. Moore, Catherine M. Martin, Richard Garrison

99 Triad Case Studies Lessons Learned from Implementation at Diverse Sites
Sheri L. Moore, Kira Lynch, Kymberly Takasaki

Session 179 - Performance Based Management, Intergovernmental Data Quality Framework and Remedial Process Optimization (RPO) Part I

Cochairs: Javier Santillan
Jeffery Perl

100 Performance Based Management Concepts and Implementation
Javier Santillan

101 Intergovernmental Data Quality Framework, Mir, Ufp
Edward Brown

Session 219 - Performance Based Management, Intergovernmental Data Quality Framework and Remedial Process Optimization (RPO) Part II

Cochairs: Javier Santillan
Jeffery Perl

102 Rpo Implementation at Afcee
John Gillespie

Session 254 - Staged Electronic Data Deliverable (SEDD) and Rapid Site Characterization Part I

Cochairs: Javier Santillan
Jeffery Perl

149 Staged Electronic Data Deliverable Short Course
Anand Mudambi

Session 267 - Staged Electronic Data Deliverable (SEDD) and Rapid Site Characterization Part II

Cochairs: Javier Santillan
Jeffery Perl

25a Innovative Environmental Management Procedures - Introduction

Jeffery Perl, Chicago Chem Consultants, 14 North Peoria, Suite 2C, Chicago, IL 60607

This time will provide an introduction to the presentations and short courses to be provided over the next three days.

The presentations and courses will provide participants with an understanding of how new technologies and approaches have been integrated with environmental projects to streamline cleanups and save money while protecting the environment in a risk protective -cost effective manner.

25b Natural Infrastructure Capabilities Requirements Management

Joann Socash, Brooks Air Force Base, 3300 Sidney Brooks, Brooks City-Base, TX 78235-5112
Military and industrial installations were traditionally located well away from population centers. In the last 30 years, however, population growth has moved the public into closer proximity to military installations (natural infrastructure shrinking), and the growth of environmental laws and regulations has imposed restrictions on what can be done on and around these installations (regulations increasing). NICRM identifies and quantifies infrastructure capability to support the military mission, thus enabling a logical, planned approach to encroachment prevention. Environmental management's focus has changed from being compliance oriented to focusing on mission needs (i.e., directly supporting operational capability) with compliance inherent.

The paper will be presented by a representative of the Office of the Secretary of the Air Force Installations and Environment.

25c Innovative Procedures to Accelerate Cleanup

Joann Socash, Brooks Air Force Base, 3300 Sidney Brooks, Brooks City-Base, TX 78235-5112
This paper will focus of the variety of procedures currently available for accelerating cleanup projects.

The invited speaker is Mr. Walt Kovalick from the Environmental Protection Agency, Office of Solid Waste.

25d Peformance-Based Management and Remediation Process Optimization Technical Regulatory Guidance Development by the Interstate Technology Regulatory Council

Thomas K. O'Neill, Bureau of Operations, Maintenance & Monitoring, New Jersey Department of Environmental Protection, PO Box 413, 401 E. State St., Trenton, NJ 08625-0413

Both Performance-based Management (PBM) and Remediation process Optimization (RPO) offer opportunities to improve the site remediation process. The Interstate Technology regulatory Council's (ITRC's) process of developing Technical Regulatory Guidance Documents, training, and a network of experts and informed practitioners expedite the process. This presentation will describe both the ITRC process and the RPO Teams findings on PBM and RPO. PBM is a strategic, goal oriented uncertainty management methodology that is implemented through effective planning and timely and documented decision logic that is focused on the end results. While use of PBM has been used in the industrial, commercial, and service sectors for many years, the use of PBM in hazardous site restoration is a relatively new development. The ITRC RPO Team is endeavoring to identify the various forms that PBM can take. The team will present, in the form a Technical Regulatory Guidance Document, to the state regulatory community, and those who interact with them (engineers, site owners, stakeholders, consultants, and the development community) the benefits and potential drawbacks of PBM.

A team of state regulator, federal partners, industry, engineering, consulting, and stakeholder representatives prepares an ITRC Technical Regulatory Guidance Document. The document development process will be described form issue identification, through development, and to the eventual development of Internet based training. The challenges and benefits of this team process will be identified. The particular issues raised by PBM will be described. Team findings, to date, as to the nature of existing PBM programs, program mandates to implement PBM, and the challenges to that implementation will be discussed. Findings from the survey of state interest in the topic of PBM will be reported.

In addition, the relationship of Remediation Process Optimization, the subject of the ITRC RPO Team's first Technical Regulatory Guidance Document, Remediation Process Optimization: Identifying Opportunities for Enhanced and More Efficient Site Remediation (RPO-1), ITRC September 2004, to PBM will be discussed. RPO is the systematic evaluation and enhancement of site remediation processes to ensure that human health and the environment are being protected over the long term at minimum risk and cost. The main difference between RPO and PBM being RPO is a review of site remediation processes and PBM is project management methodology. How the ITRC Team came to the topic of PBM while studying RPO will be discussed.

Performance-Based Management Implementation in South Carolina

Sriram Madabhushi, Hydrogeology Division, **Kent Coleman**, UST Program, and **David Scaturo**, CA Engineering Section,

Bureau of Land and Waste Management, South Carolina Department of Health and Environmental Control, 2600 Bull Street, Columbia, SC 29201

Performance-based Management (PBM) - a strategic, goal-oriented uncertainty management methodology that is implemented through effective planning and timely decision-logic that focuses on the desired end results – has been used in various forms within the Bureau of Land and Waste Management programs.

In the Underground Storage Tank program, PBM in the form of pay-for-performance (PFP) has been successfully applied for over eight years. As the number of sites with PFPs increased, the average cleanup cost has been reduced to about half of the original time and materials costs and the time to completion has also reduced considerably. Upfront fixed-price cost commitments has been beneficial in realistic budget estimates and planning. Simplified payment structure based on meeting certain intermediate goals has made payments to the contractors timely and efficient.

In the Resource Conservation and Recovery Act (RCRA) program of the Bureau of Land and Waste Management of the South Carolina Department of Health and Environmental Control, cleanup at a few contaminated DoD sites have been managed through performance-based contracts (PBCs). At these sites a general progress towards more efficient and faster cleanups is being accomplished as a result of PBCs. In the case of Charleston Naval Complex, there were about 170 sites with a variety of contaminants in soil and groundwater. Successful implementation of PBC has helped in putting corrective measures in place for most of these sites and transferring all the property to private ownership. Regulatory involvement and cooperation were main elements of this success as a large number of documents were reviewed in a relatively short period of time and prompt decisions were made in an expedited manner. More PBCs are currently being successfully implemented at Ft. Jackson, Charleston Air Force Base and Shaw Air Force Base.

Introduction: This paper summarizes several aspects of Performance-based Management (PBM) as applied in South Carolina. We try to highlight some of the successes we had in our Underground Storage Tanks (UST) Program and Resource Conservation and Recovery Act (RCRA) Programs.

UST Program – Pay –Per-Performance: South Carolina is a state with vast experience in the implementation of PBM for almost a decade. In the UST program the PBM in the form of pay-for-performance (PFP) has been applied since 1997 and during these years it evolved into a comprehensive program successfully reaching numerous milestones. There are over 4,000 reported releases more than half of which are still open releases. Of these as many as 400 releases were addressed with active corrective actions administered under the State Underground Petroleum Environmental Resource Bank (SUPERB) Act. SUPERB receives about $12-$15 million dollars yearly. Using the risk-based approach the releases are classified into different classes and the most contaminated sites which rank high in the classification are addressed on a priority basis.

The assessments or remedial investigations aimed at properly defining the contamination problems are addressed using a tiered approach. Tier I assessment involves a standard plan where a limited assessment is performed at most of the sites in order to understand basic information specific to a site. This information includes: receptor & utilities survey, site survey, installation of eight soil borings at strategic locations, installation of three monitoring wells, soil and ground water samples for laboratory analysis, aquifer evaluation using slug tests, etc. All the information gathered using the field investigations are presented in a standard reporting form.

Tier II assessment involves a more site-specific approach in which the complete characterization of the site is accomplished using primarily field screening tools followed by installation of optimal number of permanent monitoring wells. Once all the information - such as the extent of the contamination in soil and groundwater, all the exposure pathways, site geology and hydrology and the fate and transport of the chemicals of concern, etc. – specific to a site is available, then the release is classified. Based on the classification a decision for active remediation is made and the cleanup is initiated based on a pay-for-performance basis.

Based on a procurement process that awards a contract price equal to the lowest bid, corrective action is initiated for each release. Based on meeting certain goals during the remediation process, contract payments are approved. Figure 1 gives an example of how this pay for performance is approved.

Example for $100,000 Award:

Base Concentration:	11,000 ppb
Clean-up Goal:	- 1,000 ppb
Clean-up Mass:	10,000 ppb
40% Initial Money:	$40,000 for System Startup
10% for 1/4 Reduction:	$10,000 for Reaching 8,500 ppb
10% for 1/2 Reduction:	$10,000 for Reaching 6,000 ppb
15% for 3/4 Reduction:	$15,000 for Reaching 3,500 ppb
25% Completion:	$25,000 for Obtaining 1,000 ppb

Figure 1. Example payment schedule upon reaching several intermediate goals during an active corrective action process.

To ensure high quality of the performance, certain steps are utilized in the management of the these sites. A high priority is given to ensure that the site is adequately assessed with final active corrective action goals in mind from the beginning of the process. Field inspections are conducted in a regular basis and monitoring reports are reviewed on a quarterly basis. Upon reaching cleanup goals, verification monitoring is conducted at locations determined by the project managers and split samples are used to assure quality measurements.

Time and Material Average Cost	
Initial System & Startup	$150,000
O&M 5 years @ $35,000/year	$175,000
Total	$325,000

Figure 2. Average corrective action costs using previous time and material average costs.

During these eight years of PFP implementation, many benefits are readily observed. The cleanup completion times are below initially estimated times resulting in quicker cleanups. Compared to previously used 'time and materials' cost estimates, the PFP costs are almost reduced to one-half or less. Total active corrective action cost is fixed and is very helpful for upfront budgets and planning. Equipment hassles are eliminated and the contract payment process is simplified and timely. Average processing time for an invoice is a few days. Figure 2 shows average costs for earlier time and materials method and Table 1 shows how the average costs for active corrective action bids are shown. The savings from this process totals to several millions of dollars over the eight years.

Table 1. Average bid costs for the active corrective actions performed at several sites during the eight year period from 1997-2004.

Year	# Cleanups	Average Bid Cost
1997	16	$128,396
1998	68	$154,880
1999	43	$112,404
2000	25	$103,411
2001	47	$138,758
2002	62	$84,187
2003	48	$102,208
2004	52	$102,520

59b Exit Strategy and Systematic Planning within Performance Based Management

Patricia Reyes, Mitretek Systems, 3150 Fairview Park Drive South, Falls Church, VA 22042
A performance-based exit strategy focuses on performance (i.e., progress toward achieving RAOs) in order to routinely optimize the selected remedy and RAOs as information improves. Such exit strategies are dynamic and explicitly incorporate the flexibility needed to refine the strategy as site and technical knowledge improve over time, and emphasize assessment and optimization of remedy performance to assure timely, cost-efficient protection of human health.

This paper will focus on how an exit strategy is effectively developed and executed through the PBM approach and inherent systematic project planning.

59c Implementation of Performance Based Management and Performance Based Contracting

Erica Becvar, Air Force Center for Environmental Excellence, 3300 Sidney, Brooks City, TX 8235-5112
The AFCEE Technology Transfer Outreach Office developed Restoration PBM Draft Guidance that was successfully tested at DoD sites. PBM restoration incorporates, in its framework, well recognized and accepted business practices and value engineering techniques such as streamlined or accelerated site characterization (i.e, Triad), Remedial Process Optimization (RPO), development of a definitive exit strategy based on attainable Applicable or Relevant and Appropriate Requirements (ARARs), and risk management based on future land use. PBM also promotes performance-based restoration, including munitions and low-level radiological waste, by procuring services through performance-based contracts.

This presentation will provide an overview of PBM and case studies where PBM has been implemented.

IMPLEMENTATION OF THE METHODS INNOVATION RULE (MIR) AS A MAJOR FACTOR IN THE RCRA PERFORMANCE BASED MEASUREMENT (PBMS) PROGRAM

Barry Lesnik

U.S. Environmental Protection Agency
Office of Solid Waste – Methods Team

Major Topics to Be Covered

- What is the Methods Innovation Rule (MIR)?
- What will the MIR do and not do?
- How will it affect the existing RCRA Program?
- What are the responsibilities of Regulators and the Regulated Community under the Performance Approach?
- What are "appropriate methods"?

Some IUPAC Definitions (Ref. 1)

- Rational Method - a method that determines one or more identifiable chemicals or analytes for which there may be several equivalent methods of analysis availablle
- Empirical Method – a method that determines a value that can be arrived at only in terms of the method *per se* and serves, by definition, as the only method for establishing the measurand. (used to measure "method-defined parameters" or MDPs)

What is the MIR?

- The Methods Innovation Rule, or MIR, was designed to remove the unnecessary requirements to use only SW-846 methods for RCRA applications that did not require the use of empirical methods.
- The MIR allows more flexibility in RCRA-related sampling and analysis using rational methods, which is one of the Agency's steps towards implementing the Performance Approach (formerly PBMS).
- The MIR was published in the *Federal Register* as a Proposed Rule on October 30, 2002 (67 *FR* 66251).
- The MIR was published in the *Federal Register* as a Final Rule on June 14, 2005 (70 *FR* 34537).

What Will the Final MIR Do?

- Removes unnecessary required uses of SW-846 methods in the RCRA regulations and returns SW-846 to its originally intended purpose, as a guidance document
- Finalizes SW-846 Update IIIB
- Allows the Agency to fully implement the Performance Approach in the Federal RCRA Program
- Allows the Agency to publish rational SW-846 methods as guidance rather than through the formal Rulemaking process

Why is EPA Removing Unnecessary Required Uses of SW-846?

- EPA wants to allow more flexibility in RCRA-related sampling and analysis, and concentrate on measurement objectives rather than measurement technologies, i.e., the Performance Approach.

- Some regulated entities told EPA that they would like to use other reliable methods during RCRA compliance.

- Therefore, EPA has finalized the MIR to remove unnecessary requirements to use SW-846 methods in the RCRA regulations, and allow the use of other appropriate methods.

What Required Uses of SW-846 Methods Are Necessary and Have Not Been Removed?

- The MIR did not remove requirements to use a method if it is the only method capable of measuring the regulated property – i.e., empirical methods used to measure method-defined parameters (MDPs). Regulated entities will still have to use those methods.
  - For example, 261.24 (a) requires the use of SW-846 Method 1311, TCLP, for toxicity characteristic analysis
  - The required use of the TCLP is necessary and thus has not been removed by the MIR
- We identified 29 SW-846 methods that may be required to analyze RCRA-required MDPs.

What Required Uses of SW-846 Methods Are Not Necessary and Have Been Removed?

- Other required uses of SW-846 methods do not involve the analysis of required MDPs.
  - For example, § 260.22(d)(1)(i), part of the delisting regulations, requires the use of SW-846 methods to delist a waste listed with code "T".
- For these regulations other methods can be used without affecting the regulatory decision, provided the project data quality objectives (DQOs) are met.
- The required use of only SW-846 methods is not necessary, and the use of other appropriate methods is allowed.

What Might be the Impact of Removing Required Uses of SW-846?

- Regulated entities will be able to use other methods besides the SW-846 methods, when appropriate. They might choose these other methods to save costs and time.

- This greater flexibility will also stimulate the development and timely use of innovative and more cost-effective monitoring technologies and approaches in the RCRA Program.

Now That We Have Adopted the Performance Approach, What Are the Responsibilities of the Affected Entities?

- Regulatory Agency Responsibilities
  - Regulators and Permit Writers
  - Enforcement
- Regulated Community

Regulator Responsibilities Under the Performance Approach

- Regulators must establish clearly delineated DQOs appropriate for regulatory activities. Examples include:
  - Permits will need to include regulatory limits for target analytes with tolerable error rather than a general reference to use a particular method
  - Specific DQOs including tolerable error for demonstration of compliance to regulatory criteria need to be written directly into regulations.

Regulator Responsibilities Under the Performance Approach

- Enforcement approach would be to determine if the quality of the data generated and documentation provided met DQO/MQO requirements, rather than whether a particular method was followed exactly
- EPA-published methodology is guidance

Responsibilities of Regulated Community Under the Performance Approach

- Generator is responsible for demonstrating regulatory compliance
  - Selection of appropriate methods to demonstrate ability to meet project-specific DQOs/MQOs
  - Must maintain appropriate documentation for this demonstration, e.g., SAP, SOPs, QC data, etc.

Responsibilities of Regulated Community Under the Performance Approach (Contd.)

Allowable Flexibility

- Any method that yields acceptable data quality for a particular application may be used (See "Appropriate Method" slides)
- Published methods may be modified, as necessary, to generate acceptable data quality without pre-approval by EPA

Responsibilities of Regulated Community Under the Performance Approach (Contd.)

Allowable Flexibility (Contd.)

- New technology may be used as soon as it is developed and validated, provided that it can be demonstrated to be appropriate for generating acceptable data quality for a particular application
- The facility is responsible for meeting DQOs/MQOs and documenting that they were met for a particular application

How Will the Performance Approach Change the Current Way of Doing Business?

- The data review focus would shift from whether a method was exactly followed to whether the methods used were appropriate to meet the required DQOs/MQOs.
- It will require the generator to provide more substantial documentation than is currently being generated in most programs whether EPA-published methods are used or not.

What Is An "Appropriate Method"?

The MIR replaces unnecessary required uses of SW-846 by allowing the use of other "appropriate methods such as those found in SW-846 or other reliable sources."

1. Appropriate methods are reliable and accepted as such in the scientific community.
2. Appropriate methods generate effective data -- data of sufficiently known and of adequate quality for supporting the project-specific decisions.

What Is An "Appropriate Method"? (Contd.)

- Simply stated, an "appropriate method" is one that can be demonstrated to be able to see the analytes of concern at the level of concern in the matrix of concern and generate data of appropriate quality to be used for its intended application, i.e., effective data.

Who Chooses the Appropriate Method?

- Choosing an appropriate method begins during project planning and involves all key participants.
- The regulated entity is ultimately responsible for compliance with the subject regulation.
- We recommend consultation with the regulating authority during method selection, if necessary, and with the laboratory that is ultimately doing the work.

What Are the Key Steps in Choosing an Appropriate Method?

- Identify regulatory requirements (e.g., the analyte, matrix, and regulatory action level), the decision to be made, and the data quality needed to support that decision.
- Identify analytical goals (including for example performance requirements related to the matrix and analyte detection, e.g., recovery and precision).
- Select a method believed to be capable of meeting the performance requirements and other goals.

What Are the Key Steps in Choosing an Appropriate Method? (Cont.)

- Seek guidance as needed during method selection, e.g., from your laboratory, your regulating authority and SW-846 Chapter Two.

- Identify the data quality objectives, the method to be used, and the quality control indicators and make sure to put them into the planning document (e.g., QAPP, WAP, SAP).

What Are the Key Steps in Choosing an Appropriate Method? (Cont.)

- Use the data generated during the implementation phase to demonstrate that the method is appropriate – to show that the criteria were met.
- The performance data in SW-846 methods are just examples and not requirements – the actual performance acceptance criteria should be project-specific.
- As part of a dynamic approach, method selection and optimization may be revisited during the implementation phase.

Why We Selected the "Appropriate Method" Approach

1. The appropriate method approach is universally applicable to the affected regulations and requires minimal regulatory revisions.

2. The appropriate method approach is not new to RCRA regulations.

3. Also, regulated entities already should be using a similar approach to determine if a method is appropriate to a particular waste and analytical purpose.

Does the MIR Add New Requirements to the RCRA Regulations?

No new requirements are added!

- The use of <u>other</u> appropriate methods will be an option, not a requirement.
- Regulated entities can still use the SW-846 methods, if they are appropriate for the intended application.
- EPA is not adding new testing or information collection and reporting requirements beyond what should already be done now.

Adoption by Authorized State Programs

- Authorized States do not have to adopt the removal of unnecessary required uses of SW-846 methods, although we strongly encourage them to do so.
- Adoption may not cause significant impacts because States should already review method selection and quality control data to determine regulatory compliance; and most entities will probably continue to use SW-846.

EPA to Offer Training and Guidance

- EPA plans to offer training to the States and other entities on the MIR concepts and practices via training modules, workshops, and fact sheets.
- We hope that training will help ensure consistent implementation of the MIR and limit any costs.
- Please provide us with feedback as to what major topics we need to cover and which audiences we need to address

The MIR Also Finalizes SW-846 Updates IIIB

- Update IIIB includes a revised Chapter Seven, removing the reactivity test guidance.
- Update IIIB includes empirical methods used in the analysis of MDPs, and most were revised to remove the unnecessary required use of SW-846 Chapter Nine.

Status of the RCRA Waste Sampling Draft Technical Guidance

- The RCRA Waste Sampling Draft Technical Guidance was noticed as part of the proposed MIR.
- We have currently addressed all comments and are working with OGC to determine whether there is sufficient change in content to warrant our putting it out again for public comment.

New Sampling Guidance Features

- Summary of RCRA regulatory drivers for waste sampling and testing
- Emphasis on systematic planning using the Data Quality Objectives (DQO) Process
- Sampling theory and practical methods for managing sampling error
- Sampling tools
- New sampling designs and guidance on composite sampling and "hot spot" detection

New Sampling Guidance Features
(Cont.)

- Subsampling techniques
- Statistical methods and statistical tables
- Strategies for sampling heterogeneous waste
- Comprehensive examples that integrate use of the DQO Process, sampling theory, PBMS, and data quality assessment

Reference

1. *Thompson et. al., IUPAC Technical Report Pure & Applied Chemistry*, Vol. 71, No. 2, pp. 337-348, 1999.

Acknowledgement

- Thanks to Kim Kirkland of the Methods Team who completed the MIR project.

FOR MORE INFORMATION

- Contact Information for Barry Lesnik:
  - Phone: (703) 308-0476
  - E-mail: lesnik.barry@epa.gov
- Go to the RCRA methods web site at www.epa.gov/SW-846 for access to the MIR and for more information about SW-846 and the Performance Approach.
- Or contact the MICE Service for answers to your MIR or SW-846 questions at:
  - (703) 676-4690
  - (703) 318-4682
  - mice@saic.com
  - http://www.epa.gov/epaoswer/hazwaste/test/mice.htm

SEDD – An Overview and Status Report

Anand R. Mudambi

US EPA Analytical Services Branch, 1200 Pennsylvania Ave NW,
Mail Code 5102G, Washington, DC 20460

Author's e-mail: mudambi.anand@epa.gov: Phone: 703-603-8796

SEDD (Staged Electronic Data Deliverable) is a program neutral format for the delivery of analytical data. It supports multiple users' needs depending on the level of analytical requirements. The main advantage of SEDD is that once implemented, laboratories do not have to completely overhaul their Electronic Data Deliverable (EDD) generating systems as data requester needs become more complex, but can simply add additional elements to their current system. Using SEDD as the basis of electronic delivery of analytical data will decrease costs by reducing number of EDDs laboratories currently have to support and ease data exchange between various programs and agencies.

At the present time different levels or Stages for SEDD have been developed based on the complexity of data reporting requirements needed.

SEDD files are delivered as XML (eXtensible Markup Language) files. Sponsored by the World Wide Web Consortium (W3C), XML is license free, platform independent, final recommended standard which encapsulates structured data in text files. It is well supported by freely available third party tools.

Delivery of analytical data in the SEDD format is now a requirement in the U.S. Army Corps of Engineers' Formerly Used Defense Sites (FUDS) program and for certain U.S. Environmental Protection Agency contracts including the new Contract Laboratory Program's (CLP) Organic Statement of Work SOM 1.1. Interagency efforts are underway with the U.S. Air Force, U.S. Navy, and the Department of Energy to promote the use of SEDD.

Laboratories and LIMS vendors are gearing up to provide SEDD files as required in the upcoming contracts. Laboratories have already submitted compliant SEDD files as part of the new CLP Organic Solicitation. U.S. EPA is providing tools to assist laboratories in both creating and checking SEDD files prior to submission. Private parties are also evaluating SEDD as a basis for uniform delivery of analytical data.

ABSTRACT: Uniform Federal Policy (UFP-QS) for Implementing Environmental Quality Systems

Introduction

Purpose

The ***Uniform Federal Policy for Implementing Environmental Quality Systems (UFP-QS)*** outlines essential elements of a Quality System for management of environmental data collection and use and environmental technology programs. The UFP will serve as a high-level policy for documenting and implementing acceptable Quality Systems for Federal agencies. The UFP provides a framework to ensure that essential elements are addressed. The UFP must be used to develop a new Quality System or to evaluate the adequacy of an existing Quality System. The results of that evaluation must then be used to develop plans for correcting identified deficiencies.

The policy was developed as a joint initiative between the U.S. Environmental Protection Agency (EPA), the Department of Defense (DoD), and the Department of Energy (DOE) to resolve data quality inconsistencies and/or deficiencies to ensure that:

- Environmental data are of known and documented quality and suitable for their intended uses, and
- Environmental data collection and technology programs meet stated requirements.

Ultimately, the benefits of a consistent policy for Quality Systems across Federal agencies include:

- Improved effectiveness of Federal environmental programs by focusing on results, quality of data and services, and customer satisfaction;
- Clarification of roles and responsibilities in managing and overseeing environmental data and environmental technology programs;
- Sufficient confidence in the systems such that duplication of oversight efforts are minimized; and
- Enhanced accountability and public confidence in environmental decisions.

Scope

This document provides requirements and guidelines to Federal agencies for documenting and implementing Quality Systems for management of environmental data collection and use and environmental technology programs. It becomes the policy for an agency when formally adopted by that agency; each agency will determine how best to implement the policy.

This document represents a voluntary consensus policy. Implementation is, therefore, not subject to oversight by another Federal agency or to a Notice of Violation if one agency fails to implement all or part of the policy.

Background

In 1997, an audit report from the EPA Inspector General (Audit Report No. E1 SKB6-09-004107100132) examined laboratory data quality at Federal facility National Priorities List

Superfund sites. An audit report from the DoD Office of the Inspector General (OIG Report 97-098, *Laboratory Support Service for Environmental Testing*, February 21, 1997) addressed similar issues. These reports found that real or perceived inconsistencies and deficiencies in data quality within and across governmental organizations had resulted in greater costs, time delays, and the potential for increased risk.

In response to these and other audit reports, as well as to environmental data quality issues related to Federal facilities in general, EPA established the Intergovernmental Data Quality Task Force (IDQTF), chaired by the Director of the Federal Facilities Restoration and Reuse Office (FFRRO). The IDQTF operates as a partnership, reaching decisions through consensus. In addition to the chair, other consensus members of the task force include:

- The Department of Defense, represented by the DoD Environmental Data Quality Workgroup (EDQW) chairperson;
- The Department of Energy, represented by the Office of Safety, Health and Security (EM-5);
- The EPA Quality Staff, Office of Environmental Information;
- The EPA Offices of Emergency and Remedial Response and Solid Waste in the Office of Solid Waste and Emergency Response; and
- EPA Regional Offices, represented by quality assurance staff from EPA Regions 1, 2, and 8 and Federal Facility Program staff in Region 5.

DoD and DOE entered into separate MOUs with EPA. These MOUs have the following goals:

- The development of a written consensus agreement on what constitutes an adequate Quality System for environmental data collection, and
- The development of a consensus agreement that outlines the roles and responsibilities of EPA and Federal agencies with regard to data management.

The parties agreed to meet these goals first by developing them in the context of hazardous waste program management with the understanding that this will establish a framework for further work under other environmental media programs.

Basis for the Intergovernmental Quality System

ANSI/ASQC E4 was selected as the basis for the intergovernmental Quality System because it is a national standard that specifically addresses environmental data collection and use and environmental technology programs. The E4 standard is written in three parts:

- Part A addresses the management elements of a Quality System,
- Part B addresses project-specific requirements related to the collection and evaluation of environmental data, and
- Part C addresses the Quality System issues related to "design, construction and operation of environmental technology."

- The UFP is designed to clarify and provide additional direction for the implementation requirements of ANSI/ASQC E4 Part A.

The IDQTF is not developing a project-specific work product to address ANSI/ASQC E4 Part C: *Design, Construction, and Operation of Environmental Technology*.

To satisfy the second goal of the MOU, the IDQTF is developing an additional document, *Roles and Responsibilities Framework for Federal Facility Oversight.*

Documentation of Intergovernmental Quality Systems

A Quality System is documented at an organizational level in a Quality Management Plan (QMP) and at a program or project level in a Quality Assurance Project Plan (QAPP), or their functional equivalents. The organizational QMP will detail how the Quality System is to be implemented throughout the organization for which it is written. It will include information by which the organization will manage, plan, implement, assess, conduct corrective action upon, and continually improve the products, services, and activities involved in environmental data collection or use or environmental technology management.

Organization

The UFP-QS is organized according to the 10 elements that must be addressed in an organization's QMP in order to conform with ANSI/ASQC E4 and this document:

1. Management and Organization
2. Quality System and Description
3. Personnel Qualification and Training
4. Procurement of Products, Services, and Activities
5. Documents and Records
6. Computer Hardware and Software
7. Planning
8. Management of Work Process Implementation
9. Assessment and Response
10. Quality Improvement

Each of these elements is covered in a separate section of the basic document.

Requirements and Guidance

The requirements described in this policy are intended to facilitate implementation of ANSI/ASQC E4 with regard to environmental data collection and use and environmental technology development and implementation. ANSI/ASQC E4 covers both mandatory specifications and nonmandatory guidelines for Quality Systems. **In accordance with E4, this document addresses both requirements and guidance. Requirements are presented in the text that makes up the bulk of this document, while guidance is presented in text boxes. The text boxes attempt to illuminate the requirements of this policy and enhance the reader's understanding. The text boxes are not part of the formal policy.** The requirements of this document also incorporate by reference the definitions contained in the E4 standard.

Relationship to Other Standards and Policies

Questions have been raised by reviewers of this document as to the relationship between this document, international standards, and EPA policy documents. The text below is provided to describe those relationships.

Relationship to International Standards

The relationship of this policy to international standards is as follows:

- ANSI/ASQC E4 conforms with the **ISO 9000** series of standards, "International Standards for Quality Management," addressing Quality Systems. E4 and this policy document, however, are specifically written for environmental data collection and environmental technology programs.
- The **ISO 14000** series, *Environmental Management* addresses what an organization does to minimize harmful effects on the environment caused by its products, services, and activities. ISO 14000 does not address the issues covered by E4 or by the ISO 9000 series.
- **ISO/IEC Guide 25** (superceded in 1999 by ISO 17025, *General Requirements for the Competence of Testing and Calibration Laboratories*) establishes specifications and requirements for calibration and testing of laboratory systems. ISO 25 is the basis for standards set by the National Environmental Laboratory Accreditation Conference, Chapter 5. While E4 does not explicitly address quality systems for environmental laboratories, it references ISO/IEC Guide 25, and other national standards as the framework for other Quality Systems.

Relationship to EPA Policy

ANSI/ASQC E4 sets forth the general standards for a Quality System for environmental data collection and technology. EPA recently published a number of documents designed to implement E4 throughout EPA and for EPA funded activities. These documents include:

1. *Policy and Program Requirements for the Mandatory Agencywide Quality System* (EPA Order 5360.1 A2),
2. *EPA Quality Manual for Environmental Programs* (EPA Manual 5360 A1), and
3. *EPA Requirements for Quality Management Plans* (EPA QA/R2), which establishes Quality System requirements for external organizations that implement programs for EPA through funding mechanisms (e.g., contracts, grants, and other financial agreements), or through other agreements (e.g., interagency agreements).

The EPA Order and the EPA Manual apply solely to EPA. This document differs from E4 and from QA/R2 in four important ways:

- It provides implementation policies for Federal agencies; therefore, it is more detailed than either E4 or QA/R2.
- It is intended to apply to Federal agencies, whereas QA/R2 applies only to organizations that implement programs for EPA.
- It represents a consensus policy reflecting the views of the IDQTF,
- It benefits from the experience and lessons learned from a broader set of stakeholders involved in managing quality across several Federal agencies.

This document is consistent with the EPA Order and EPA Manual. It is anticipated that Quality Management Plans (QMPs) that meet the requirements of the EPA Order will require little adjustment to conform to this policy.

- The term "Quality System," as used in this document, is adopted from ANSI/ASQC E4, ***Specifications and Guidelines for Quality Systems for Environmental Data Collection and Environmental Technology Programs***. ANSI/ASQC E4 states that a

Quality System is *"A structured and documented management system describing the policies, objectives, principles, organizational authority, responsibilities, accountability, and implementation plan of an organization for ensuring quality in its work processes, products (items), and services. The Quality System provides the framework for planning, implementing, and assessing the work performed by an organization and for carrying out required quality assurance (QA) and quality control (QC) activities."*

- Whenever the term "Federal agencies" is used in this document, it refers to agencies, departments, and instrumentalities of the United States.
- Other EPA offices and DoD components participated in various IDQTF subgroups and meetings to provide technical expertise. In addition to the above consensus members, participation of various other Federal agencies and EPA offices was actively sought throughout the development of initial IDQTF products. These included the Departments of Agriculture, Commerce (National Oceanographic and Atmospheric Administration), and Interior (U.S. Geological Survey), who were invited to become consensus members.

Reference: http://www.epa.gov/fedfac/documents/data_quality/ufp_sep00_intro.htm

HEADQUARTERS U.S. AIR FORCE

The Air Force Center for Environmental Excellence

The Uniform Federal Policy for Implementing Environmental Quality Systems (UFP-QS)

Ed Brown
AFCEE/TDE

Integrity - Service - Excellence

As of: 23 Sept 05

Key Questions

- I. What is the UFP-QS?
- II. What is a Quality System?
- III. What is the relationship between UFP and EMS?
- IV. What is the UFP-QS Approach?
- V. How to access the UFP-QS document?

I. What is the UFP-QS?

- The UFP-QS is a consensus document that addresses real and perceived inconsistencies and deficiencies in quality control for laboratory data within and across governmental organizations which resulted in greater costs, time delay, and potential increased risk.
- The UFP-QS was developed as a collaborative effort by Intergovernmental Data Quality Task Force including DOE, EPA, and DOD in 1997.
- The UFP-QS was approved:
  - For USEPA use on November 1, 2002
  - For DOD use on January 17, 2003
  - For DOE use on December 13, 2003

UFP-QS Objectives

- Provides essential elements of a quality system for management of environmental data collection and use.
- Provides a framework to Federal agencies for documenting and implementing an acceptable intergovernmental quality system that is based on: *Part A of "Specifications and Guidelines for Quality Systems for Environmental Data Collection and Environmental Technology Programs, the American National Standards Institute/American Society for Quality Control Standard E4 (ANSI/ASQC E4, 1994)."*

UFP-QS Objectives (cont'd)

- **Provides guidance to document, assess, and improve existing quality systems.**
- **Encourages Federal agencies to adopt this document as policy, rule, or guidance for managing environmental data collection and use.**

Integrity - Service - Excellence

II. What is a Quality System?

- The Quality System is a structured and documented management system that describes the policies and implementation plan of an organization. It assures that data used to support environmental decisions are of adequate quality and usability for their intended purpose.

- The overall goal of this consensus system is simple: sound decisions must be based on sound documented data.

What is a Quality System? (cont'd)

Parts of Quality System

- <u>At organization level</u>: **It is documented in a Quality Management Plan (QMP). The QMP details information by which the organization will manage, plan, implement, assess, and continually improve the activities involved in environmental data collection and use.**

- <u>At project level</u>: **It is documented in a Quality Assurance Project Plan (QAPP). It is based on ANSI/ASQC-E4,**

  Part B, "collection and evaluation of environmental data."

III. What is the Relationship Between UFP and EMS

- **The UFP offers an implementation tool for meeting Section 5(d)(15) of the DOE O 450.1, *"Ensuring the analytical work supporting environmental monitoring is implemented using:***

  (a) a consistent system for collecting, assessing, and documenting environmental data of known and documented quality."

- **The UFP-QS provides "requirements and guidance on the establishment and management of quality systems for environmental testing laboratories that intend to perform work for DoD" as found in DoD's Quality Systems Manual for Environmental Laboratories, version 3.**

- **Failure to have in place a Quality System for the management of environmental data collection and use programs as part of an EMS is costly.**

Relationship Between UFP and EMS (cont'd)

- Example:

(i) lack of procedures to store and retrieve previously collected environmental data often creates a need for the collection of new data.

(ii) Failure to implement a systematic planning process often leads to the collection of wrong data at a quality not appropriate for its intended use. This creates a need for additional collection and analysis of samples.

IV. What is The UFP Approach?

SUPPORTS SOUND EMS IMPLEMENTATION

Systematic Planning:

Based on scientific method, common sense, and a graded approach to ensure that the level of detail in planning depends on the importance and intended use of the work and available resources.

Holistic Approach:

Consolidates site-wide data collection across projects, programs, and media. It enhances knowledge, communication, and promotes a graded approach necessary to optimize and conserve valuable resources.

Maintaining Linkage:

Maintains linkage among systematic planning, implementation, and assessment stages of environmental data collection to ensure that the data produced meet their intended purpose and are documented and defensible.

As of: 23 Sept 05 *Integrity - Service - Excellence* 10

Questions?
Integrity - Service - Excellence
As of: 23 Sept 05
11

V. How to access the UFP-QS?

- **The UFP-QS is available at:**
  http://www.epa.gov/swerffrr/library/data.htm.

- **ANSI/ASQC E4 is available at:**
  ASQC, P.O. Box 3005, Milwaukee, WI 53201
  (414) 272-8575

- **AFCEE Contact:**
  Ed Brown at: edward.brown@brooks.af.mil
  (210) 536-5665

Integrity - Service - Excellence

59g Performance Based Management Panel Discussion

Javier Santillan, Technical Directorate, Environmental Science Division, HQ Air Force Center for Environmental Excellence, 3300 Sidney Brooks Rd, Brooks City-Base, TX 78235-5112
The PBM Panel Discussion is intended to provide an open forum for questions, comments, and discussion of any aspect of Performance Based Management.

The panel will include the following: Tom O'Neil, Sriram Madabhushi, Deanna Crumbling, Jeff Perl, Barry Lesnik, Anand Mudambi, Ed Brown, and Javier Santillan.

American Institute of Chemical Engineers
Spring Meeting
Orlando FL
April 2006

Fixing Dysfunctional Remediation Teams

Jeffery P. Perl, PhD, PE, CHMM

© Jeffery P. Perl and Chicago Chem Consultants Corp 2006 1

Quality Management

Three Elements to Quality

1 Core Quality Management Process:

– Quality Air Force

– Total Quality Management

2 Specific System

- Systematic Planning

3 Specific Training

© Jeffery P. Perl and Chicago Chem Consultants Corp 2006 2

Generic Quality Management Model

1. Identify the Problem
2. Analyze the Facts
3. Generate Alternatives
4. Select Alternative
5. Implementation
6. Evaluation

Specific Model

ROD Compliance

Source Removal/Control

Available Treatment Options

Risk Assessment

Performance Based Measurement Systems

Systematic Planning

Cost and Schedule to Complete Review

Specific Training

RPO Team Training & Tools

© Jeffery P. Perl and Chicago Chem Consultants Corp 2006 3

Core QM Process

- Identify Affected Community
- Draw Circle Around it
- Identify Groups
- Create Team
- Set the Problem
- Solve the Problem
- Start Over

© Jeffery P. Perl and Chicago Chem Consultants Corp 2006 4

Specific Managed Program

- Remedial Process Optimization
  - DQO
  - Source Removal/Control
  - Available Treatment Options
  - Risk Assessment
  - Performance Based Measurement Systems
  - Cost and Schedule to Complete Review

© Jeffery P. Perl and Chicago Chem Consultants Corp 2006 5

Whose Side Are You On ?

Federal Regulator

Contractor

RAB

LRA

BEC

State Regulator

Team Chief

Local Regulator

PA

© Jeffery P. Perl and Chicago Chem Consultants Corp 2006 6

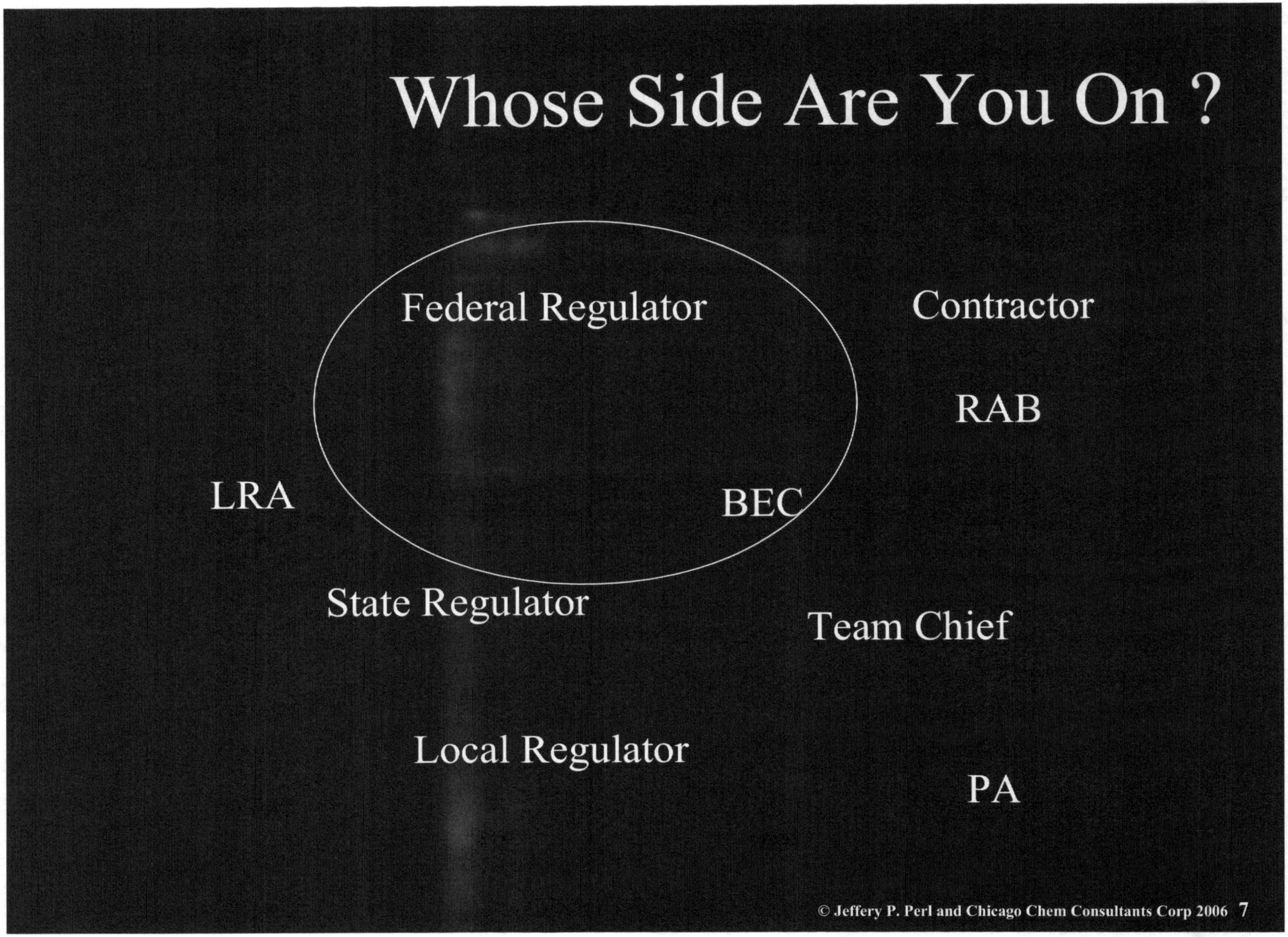
Whose Side Are You On ?
Federal Regulator
Contractor
RAB
LRA
BEC
State Regulator
Team Chief
Local Regulator
PA
© Jeffery P. Perl and Chicago Chem Consultants Corp 2006 7

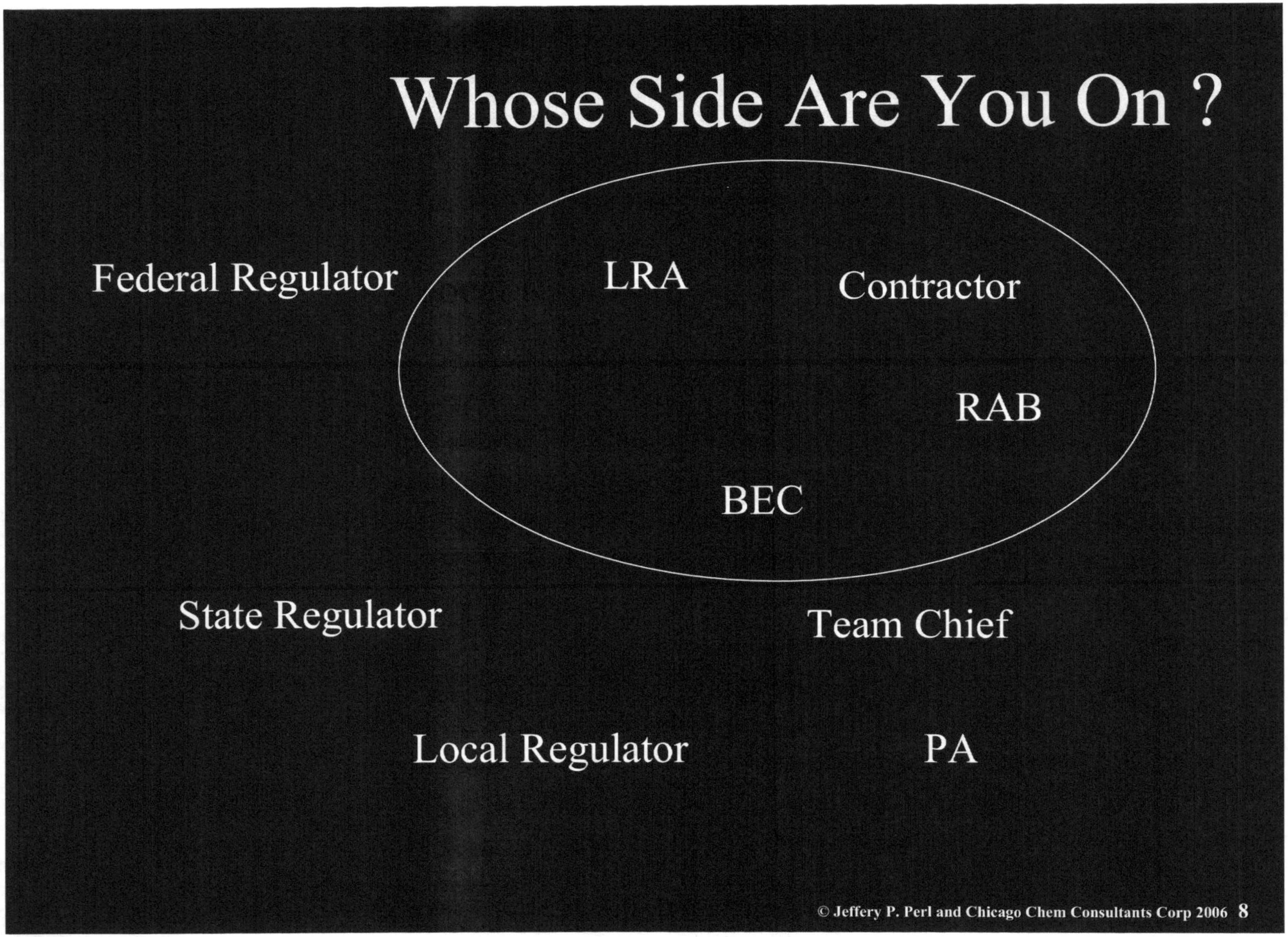
Whose Side Are You On ?
Federal Regulator
LRA
Contractor
RAB
BEC
State Regulator
Team Chief
Local Regulator
PA
© Jeffery P. Perl and Chicago Chem Consultants Corp 2006 8

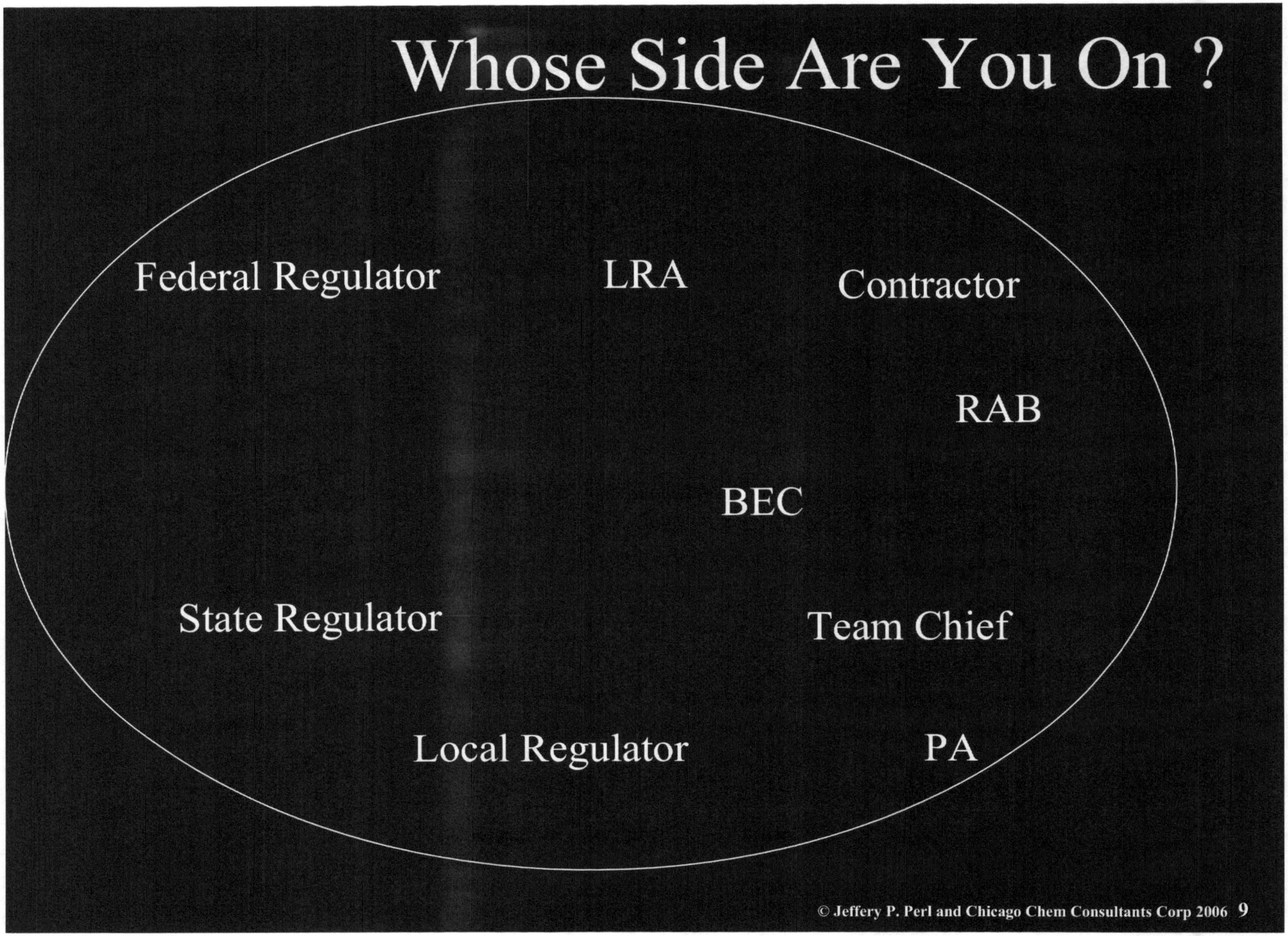
Whose Side Are You On ?
Federal Regulator
LRA
Contractor
RAB
BEC
State Regulator
Team Chief
Local Regulator
PA
© Jeffery P. Perl and Chicago Chem Consultants Corp 2006 9

Specific Training

- Geared Toward RPO
  - Team Building
  - Resource Identification and Management
  - Conflict Resolution
  - Public Speaking
  - OSHA & EPA Regulations
    - Federal/State/Local
  - AF Regulation/AFIs Etc
  - AF/EPA/Civilian Guidance & Manuals
  - Interagency Activity
  - Site Specific Community Concerns

© Jeffery P. Perl and Chicago Chem Consultants Corp 2006 10

Sample Conflicts

- Method and Manner of Risk Assessment
- Improper use of EE/CA
- Failure to Recognize Community Concerns
- Incomplete Restoration Team
- Incomplete Restoration Advisory Board
- Poor Understanding of Statutes and Regulations
- Missing or Inaccurate Communication PA

© Jeffery P. Perl and Chicago Chem Consultants Corp 2006

Sample Conflicts

- Confusing Science with Engineering
  - Know How - Science
  - Know Why - Engineering
- Failure to Identify Novel Reuse
  - Keesler Barrier Reef - P2
  - Rantoul Airport
  - Construction Debris in Landfills - P2

© Jeffery P. Perl and Chicago Chem Consultants Corp 2006 12

Conference Focus

CONTINUOUS IMPROVEMENT

NEW ERA of COOPERATION

- Promote Risk Protective/Cost Effective RA
- Recognize Imperfect Initial Site Characterization
- Incorporate New Understanding of Risk
- Cull Advances in Techniques & Technology
- Put it all Together
- Climb Through This Open Window

© Jeffery P. Perl and Chicago Chem Consultants Corp 2006 13

RAPIC SITE CHARACTERIZATION (RSC)

EXTENDED ABSTRACT

Rapid Site Characterization (RSC) is a component of Performance Based Management (PBM) which is the Air Force's philosophy and approach in managing and assessing the performance of its Environmental Restoration Program. Accelerated characterization principles were originally developed by ASTM and the Interstate Technology Regulatory Council (ITRC). Based on the ASTM and ITRC guidance, RSC has been developed by the Air Force Center for Environmental Excellence (AFCEE) to emphasize the need to incorporate evolving scientific and technical information on site characterization actions from which remediation decisions can be made, and when conducted as part of Performance-Based Management principles protectiveness can be documented and completed in a reasonable time frame in a cost effective manner. The RSC's purpose is to provide environmental decision makers and practitioners with a tool to rapidly, cost-effectively, and reliably deliver a mental picture of current environmental conditions of a site. This mental picture is conveyed through a vehicle called a representative Conceptual Site Model (CSM). The CSM is the vehicle used to succinctly and clearly identify and depict the environmental problem at a site. RSC has six interlinking components (initial CSM, Dynamic Work plans, Real Time Data Analysis, Dynamic Document Review, Decision Logic, and Systematic Planning) with the central component being systematic planning. It is through systematic planning that the other components bring together the site understanding through the representative conceptual site model. This information becomes the basis for determining the necessity of site remediation. It is during the RSC that critical parameters for the design and optimization of the remedial system are identified and measured. The RSC uses a multi-discipline team to execute the various components. This core technical team includes analytical chemists, engineers, geologists, geochemists, geophysicists, hydrogeologists, risk assessors, and regulators. This core team helps with data management, QA/QC, risk assessment, fate and transport modeling, remedial action, community relations, and health and safety. The technical team will be responsible for: 1) Gathering all available information for the site, 2) developing an initial "conceptual" model for the site, 3) identifying the technical objectives and goals to be accomplished, 4) supervising the field effort, making adjustments to the Conceptual Site Model based on the data produced in the field, and 5) evaluating the conceptual model and decisions made with respect to federal, state, and local regulations. MAJCOMs and installations may request site-specific support for accelerated or Rapid Site Characterization through the AFCEE Project Manager and/or the appropriate directorate representative letting the contracts for the project or site. General information or questions can be answered by sending an email to the AFCEE Technology Transfer Outreach Office (TTO) at afcee.td.awag@brooks.af.mil. The phone number is commercial 210-536-4366 (DSN 240-4366).

The Air Force Center for Environmental Excellence

AFCEE's Rapid Site Characterization Program

Ed Brown
AFCEE/TDE

Integrity - Service - Excellence

Outline

- Definition
  - PBM
  - RSC
- PBM Components
- RSC Components
- Purpose
- Process

PBM Definition

Performance-Based Management

- A holistic and systematic results-oriented uncertainty, contingency, and information management assessment to expedite closure through the incorporation of:
  - Clear and accurate definition of the environmental problem through updated conceptual site model (CSM)
  - Established Exit Strategy
  - Development/Implementation of Decision Analyses & Logic
  - Identification & Tracking of performance-based Metrics
  - Performance-based Contracts

PBM Definition

- Performance-based Management is a strategic goal-oriented uncertainty management methodology that is implemented through systemic planning and dynamic decision-logic focused on the desired end-results

PBM Components

ARAR	Applicable or relevant & appropriate requirements
CSM	Conceptual Site Model

Integrity - Service - Excellence

As of: 23 Sept 05

5

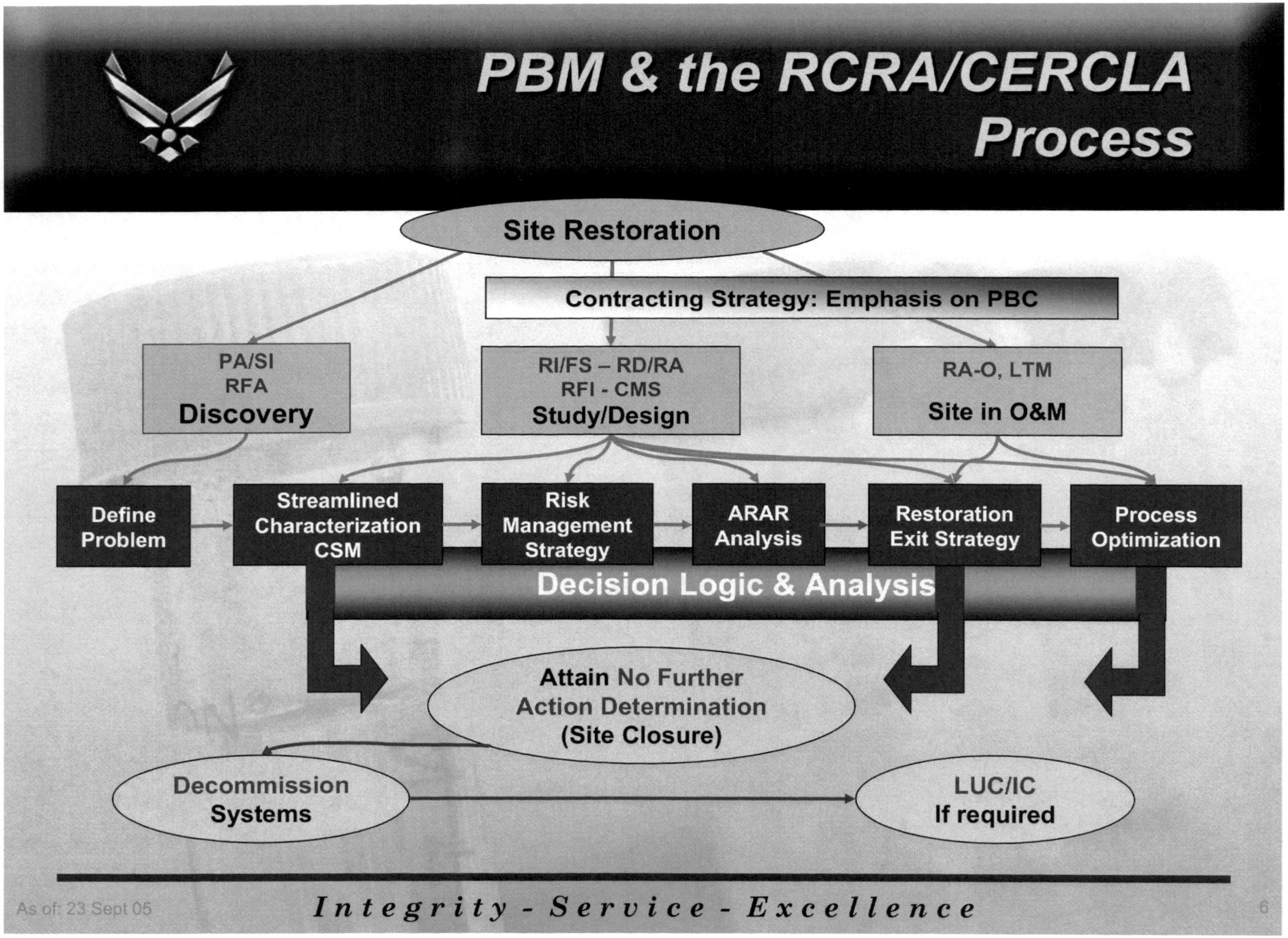
PBM & the RCRA/CERCLA Process
Site Restoration
Contracting Strategy: Emphasis on PBC
PA/SI
RFA
Discovery
RI/FS – RD/RA
RFI - CMS
Study/Design
RA-O, LTM
Site in O&M
Define Problem
Streamlined Characterization CSM
Risk Management Strategy
ARAR Analysis
Restoration Exit Strategy
Process Optimization
Decision Logic & Analysis
Attain No Further Action Determination (Site Closure)
Decommission Systems
LUC/IC If required
As of: 23 Sept 05
Integrity - Service - Excellence
6

What is Rapid Site Characterization (RSC)?

- Accelerated procedure based on systematic decision logic to delineate contamination at complex sites and construct representative hydrogeologic conceptual site models

Integrity - Service - Excellence

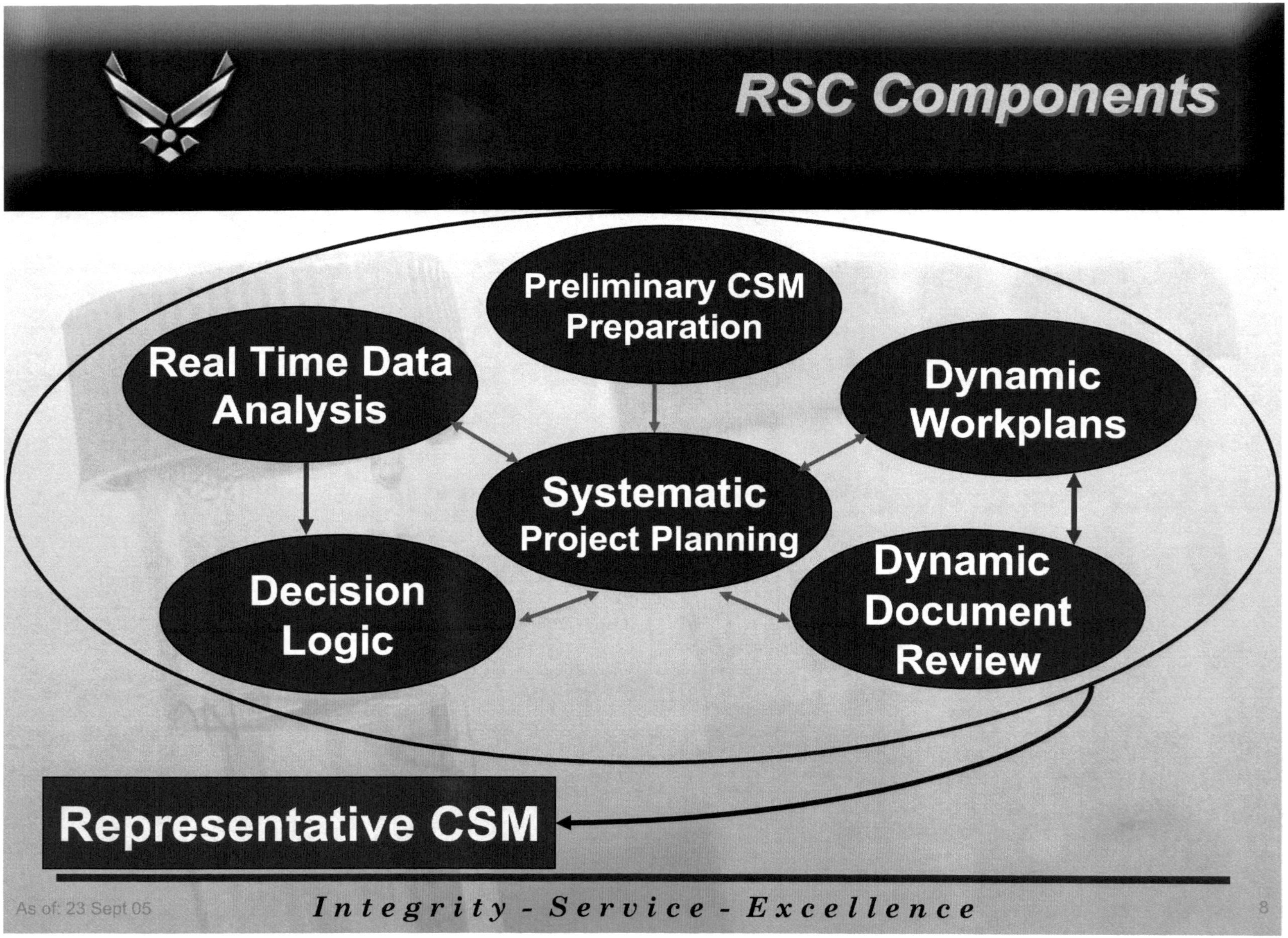
RSC Components
Preliminary CSM Preparation
Real Time Data Analysis
Dynamic Workplans
Systematic Project Planning
Decision Logic
Dynamic Document Review
Representative CSM
As of: 23 Sept 05
Integrity - Service - Excellence
8

Why use RSC?

- Create a functional restoration team
  - Face-to-face systematic planning more intensive, more far-reaching, than most imagine
- Empower field managers to make decisions
- RSC CSM captures heterogeneity & physical reality
- Analytical quality is NOT THE SAME AS Data quality
- A CSM depends on "Data representativeness"

- "Basic premise" plan before initiating field work
  - Every sample taken must lead to making a decision
  - Never take a sample that has no decision impact

Analytical Data Quality vs. Representativeness

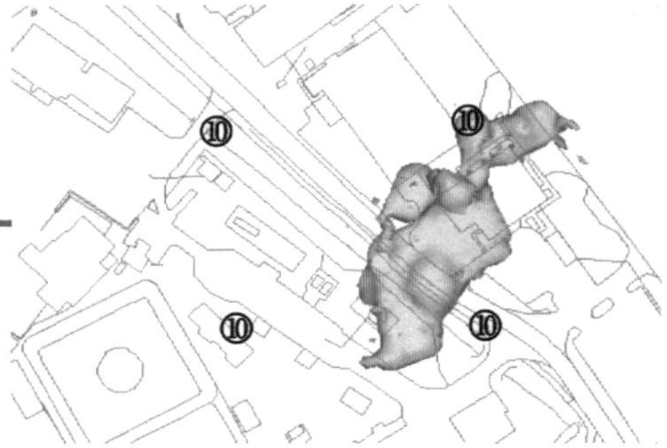

?

Four definitive quality data points generate a non-representative picture of the site

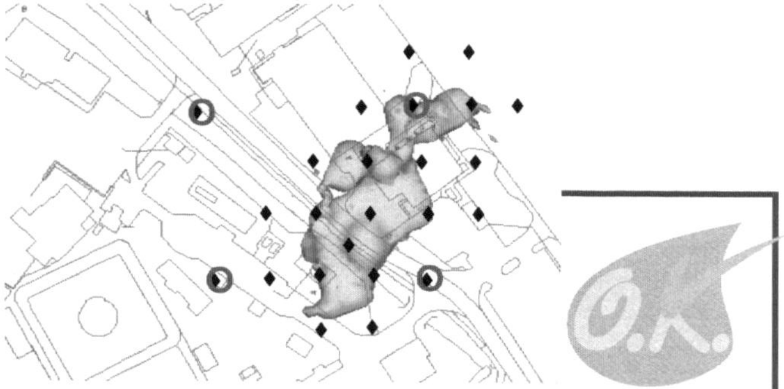

Screening quality data & sampling decision logic generates representative picture of the site

Goal: Representative Conceptual site Model

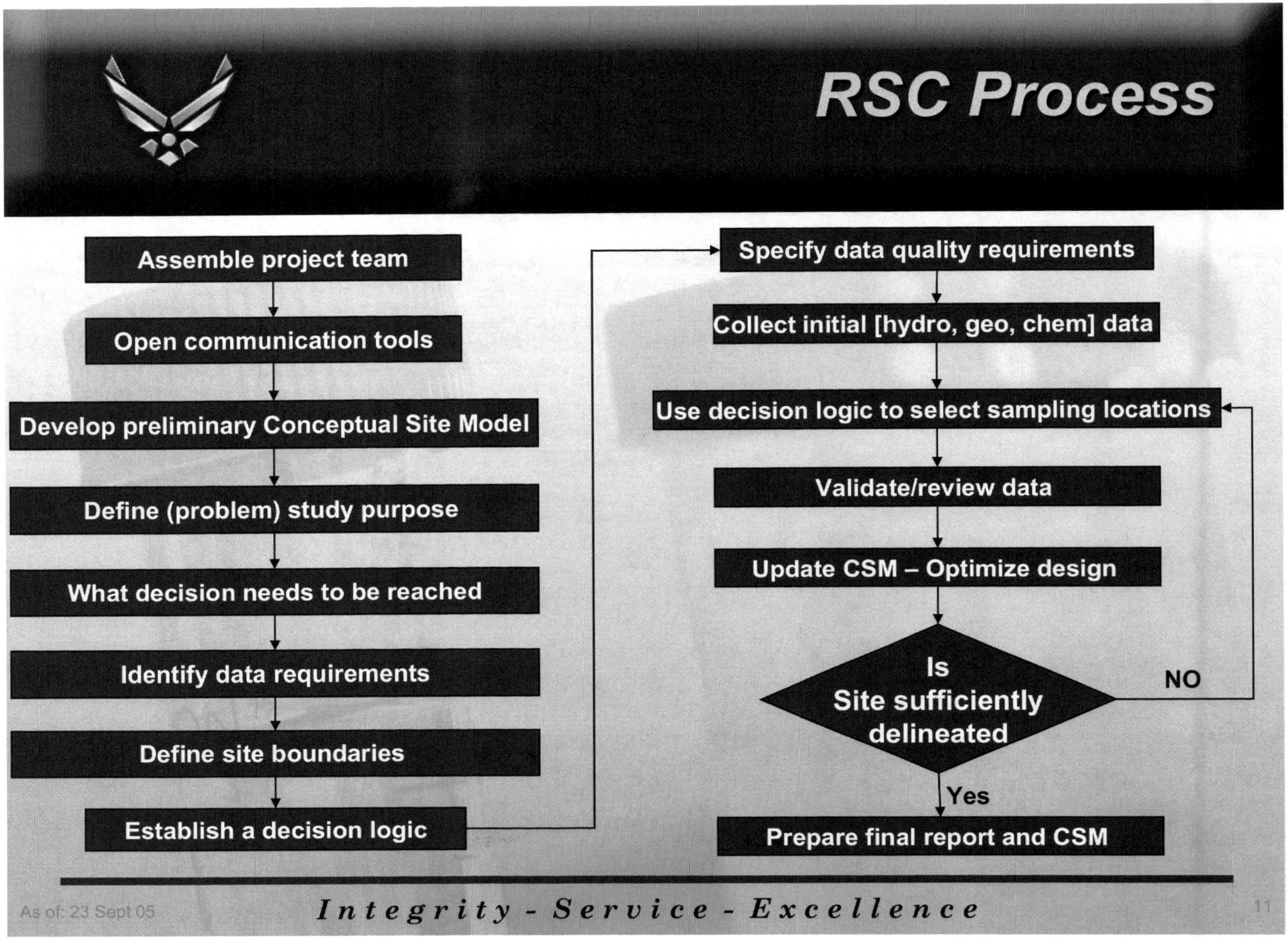
RSC Process
Assemble project team
Open communication tools
Develop preliminary Conceptual Site Model
Define (problem) study purpose
What decision needs to be reached
Identify data requirements
Define site boundaries
Establish a decision logic
Specify data quality requirements
Collect initial [hydro, geo, chem] data
Use decision logic to select sampling locations
Validate/review data
Update CSM – Optimize design
Is Site sufficiently delineated
NO
Yes
Prepare final report and CSM
As of: 23 Sept 05
Integrity - Service - Excellence
11

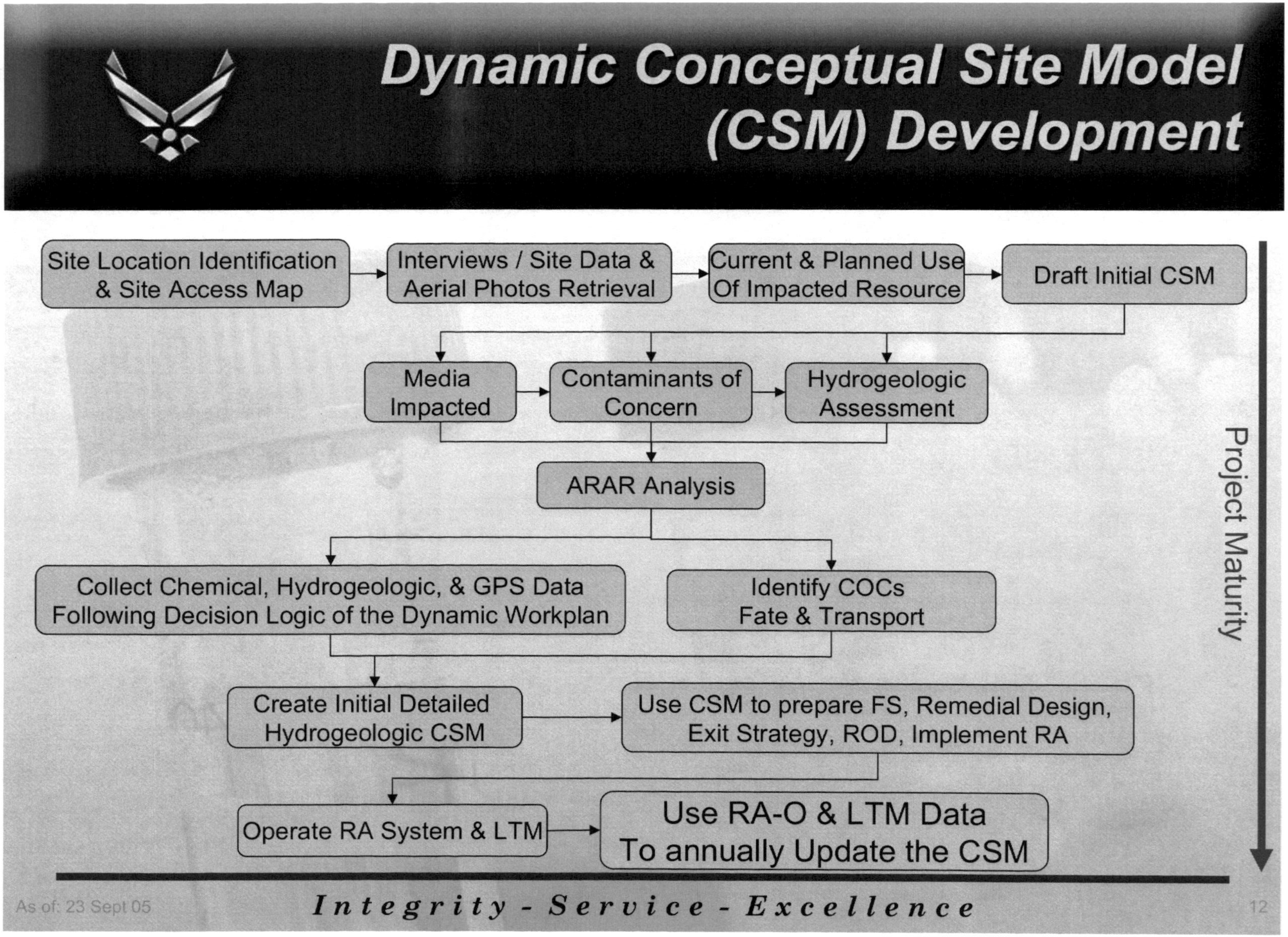
Dynamic Conceptual Site Model (CSM) Development
Site Location Identification & Site Access Map
Interviews / Site Data & Aerial Photos Retrieval
Current & Planned Use Of Impacted Resource
Draft Initial CSM
Media Impacted
Contaminants of Concern
Hydrogeologic Assessment
ARAR Analysis
Collect Chemical, Hydrogeologic, & GPS Data Following Decision Logic of the Dynamic Workplan
Identify COCs Fate & Transport
Create Initial Detailed Hydrogeologic CSM
Use CSM to prepare FS, Remedial Design, Exit Strategy, ROD, Implement RA
Operate RA System & LTM
Use RA-O & LTM Data To annually Update the CSM
Project Maturity
As of: 23 Sept 05
Integrity - Service - Excellence
12

References

- ASTM Active Standards D5730-04, E1689-95, E1912-98, and PS 3 – 95
- Regulatory Cooperation Work Group (ITRC) / American Society For Testing And Materials (ASTM) Partnership For Accelerated Characterization FY-97 Summary Report –Final December 1997

Want to Find Out More?

- Standard Guide for Accelerated Site Characterization for Confirmed or Suspected Petroleum Releases : http://www.astm.org/cgi-bin/SoftCart.exe/DATABASE.CART/REDLINE_PAGES/E1912.htm?E+mystore
- Interstate Technology & Regulatory Council TechReg Guideline for Triad: http://www.triadcentral.org/ref/ref/documents/SCM-1.pdf
- Regulatory Cooperation Work Group (ITRC) / American Society For Testing And Materials (ASTM) Partnership For Accelerated Characterization: http://www.itrcweb.org/Documents/ASC-1.pdf
- Performance Based Management and Rapid Site Characterization Guidance:

  http://www.afcee.brooks.af.mil/products/techtrans/

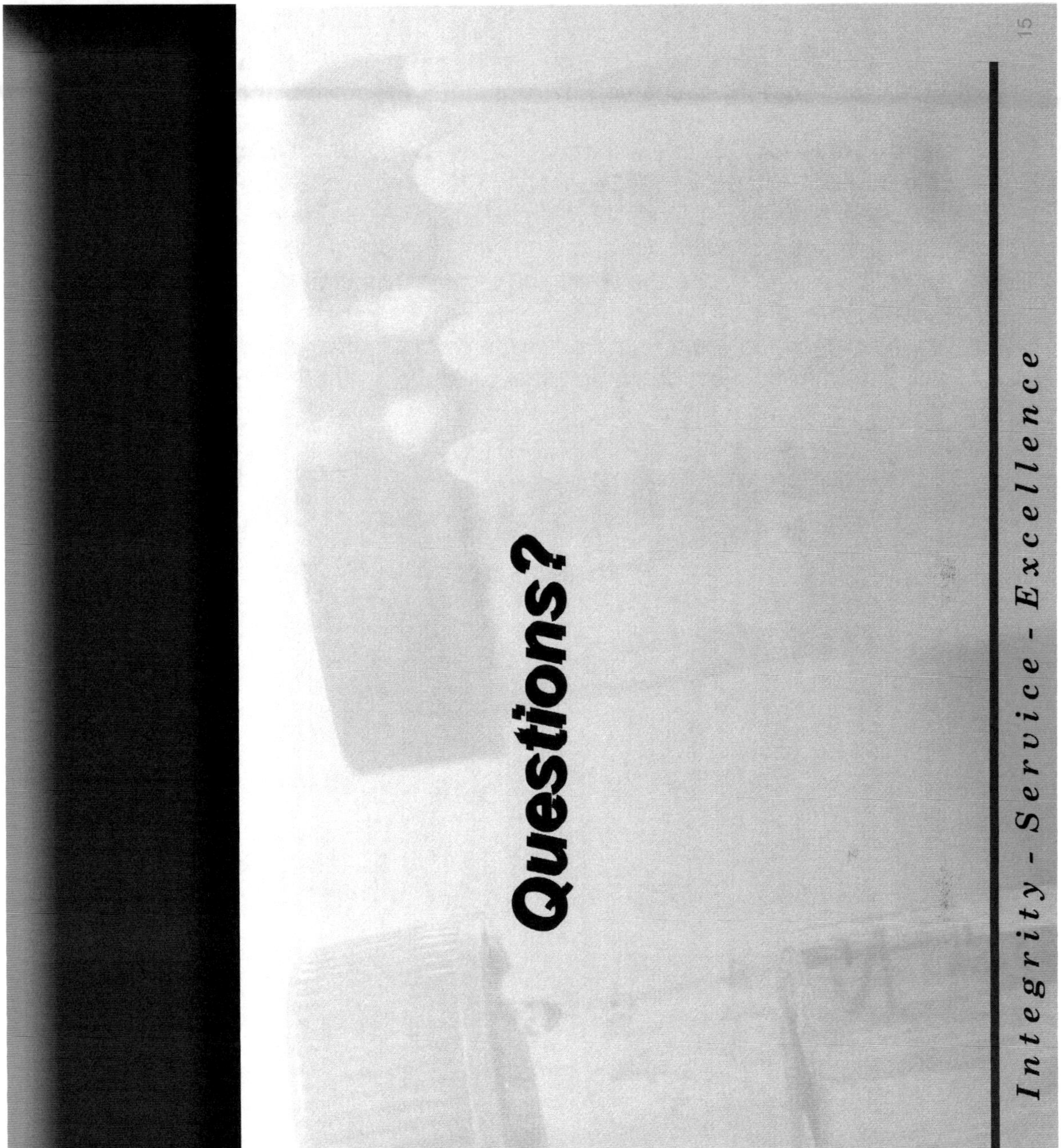
Questions?
Integrity - Service - Excellence
As of: 23 Sept 05
15

137a Concurrent Sustainable Environmental Program Management

Sekar Balusamy, Environmental, ECO-FRIEND INC, Calgary, Canada, 224 Citadel Pass Court, NW, Calgary, AB T3G 3V2, Canada
This innovative environmental restoration and environmental program management is based on proactive activities such as site remediation, site exchange plant and forest cultivation and development and other methods. Companies can plan their environmental restoration and environmental program management well before the project conceptualization stage so that the time will not run out while at the project stage.

Removal and Recovery of Heavy Metals from Industrial Waste Streams by Means of a Hybrid-Precipitation and Polymer Enhanced Ultrafiltration

Sezin İslamoglu[1], Prof.Dr.Levent Yılmaz[1], Prof.Dr. H.Önder Özbelge[1]
[1]Department of Chemical Engineering, Middle East Technical University

One of the main causes of industrial pollution is the discharge of effluents containing heavy metals. Heavy metals, especially cadmium can have serious effects on human and animal health. Beside the health effects, heavy metals are non-renewable resources. Therefore, effective recovery of heavy metals is as important as removal of them from waste streams.

Wastewater produced during electroplating is one of the main causes of contamination of the natural environment with metal ions. The negative impact of waste streams containing heavy metals upon the environment, gives rise to increasingly strict regulations. According to recent regulations of EPA, for common metals facilities discharging 38,000 liters or more process wastewaters (resulting from the process in which a ferrous or nonferrous basis material is electroplated with copper, nickel, chromium, zinc, tin, lead, cadmium, iron, aluminum, or any combination thereof) per day should obey the following limitations listed in Table 1[1].

Table 1. Common Metals Facilities Discharging 38,000 Liters or More Per Day Pretreatment Standards for Existing Sources Limitations (mg/l)

Pollutant	Maximum for any 1 day	Average of daily values for 4 consecutive monitoring days shall not exceed
Cu	4.5	2.7
Ni	4.1	2.6
Cr	7.0	4.0
Zn	4.2	2.6
Pb	0.6	0.4
Cd	1.2	0.7
Total metals	10.5	6.8

Amoung the industrial waste effluents containing heavy metals, treatment of electroplating wastes is a serious worldwide problem, because of their high content of economically valuable heavy metals.

For the fractional separation of heavy metals from electroplating waste streams, the coupling of two or more techniques may result in better performance than using either unit operation individually. By combining the membrane techniques with other physical and chemical processes the effectiveness of the operation can be improved. Precipitation of sparingly soluble metal compounds and micro-or ultrafiltration is the suitable and economical hybrid operation for the removal and recovery of heavy metals from waste streams.

Precipitation of heavy metals with lime, sulfide and caustic soda are the most commonly used economical methods for metal removal. Besides the cost effective properties, its appropriateness as a first step of a hybrid separation processes makes the precipitation widely used method. However, heavy metals can not be removed efficiently by hydroxide precipitation. Higher degrees of removal can be achieved by sulfide precipitation, but in this case high volumes of toxic sludge which are difficult to dispose are produced [2]. More sustainable, effective and economic method could be rectified for heavy metal removal and recovery if a process scheme which can achieve selective separation of metals and produce reusable pure metal sludges could be developed.

In this study, for selective removal and recovery of the metals like Cd, Cu, Fe in cadmium electroplating bath (containing high amounts of Cd, Zn, Cu, Fe and small amounts of Ni, Co, Mn), hybrid precipitation- polymer enhanced ultrafiltration based separation scheme was developed.

The precipitation scheme comprised three consecutive steps: 1)Acid treatment with nitric acid (HNO_3). 2) Alkali precipitation by sodium hydroxide (NaOH). 3) Sulfide Precipitation by sodium sulfide (Na_2S). Dosage of precipitating agent addition, pH sequencing are the parameters of the precipitation experiments.

Results of the precipitation experiments may be summarized as follows: Cyano-metal complexes were decomposed. Whole Fe content and almost half of the Ni were removed in acidification step. Then by adding sodium hydroxide heavy metals were precipitated and pure $Cd(OH)_2$, was obtained which is a valuable product used in electroplating industry. In sulfide precipitation, pH is an important parameter, such as: Addition of sodium sulfide in alkali pH range leaded to cadmium precipitation whereas copper is totally precipitated in acidic pH range.

Depending on the quality and quantity of the desired solution obtained at the end of the precipitation experiments, different acidification, alkalination and sulfide precipitation paths were followed. Cd-rich samples were obtained by making sulfide precipitation in acidic pH range where as Cu-rich samples were obtained by making sulfide precipitation in basic pH range.

After precipitation experiments, PEUF was applied. Polymer enhanced ultrafiltration (PEUF) is one of the complexation enhanced ultrafiltration processes based on the complexation between a water-soluble polymeric binding agent and the target component, which is desired to be separated from the aqueous solution by an ultrafiltration process [3]. Binding capacities of PEI with Cd, Cu, Zn and Ni strongly depend on pH. As a consequence, pH plays a significant role in the retention of heavy metals [4].

For Cu-free, Cd-rich solutions, at pH 3 and pH 4, retention of zinc and nickel were higher than the retention of cadmium. As the pH value increases, retention of Cd increases which means Cd could be removed from solution more effectively as compared to zinc and nickel.Relatively low retention in low pH values may be due to more protonation of nitrogen donor groups on the polymer at low pH.PEI exists in two forms in aqueous solutions, neutral and protonated. In neutral form free imine sites of PEI are available to form complexes with metal ions whereas in the protonated form, protons bind on PEI. Decreasing the pH enhances the binding of H^+ ions to form a complex with PEI [5].

Therefore, it can be concluded that at low pH values protons compete with Cd ions and inhibit them to bind to PEI. Zn and Ni retentions were higher than Cd retention at low pH values. Therefore it can be suggested that, at low pH values Zn and Ni ions are capable of replacing with H^+ ions to form a complex with PEI.

Two experiments were performed at pH 3 and pH 6.5 with 0.02 loading by Cu rich samples. Retentions of Cd, Zn, Ni and Cu increase when pH was increased from pH 3 to pH 6.5. Retention of Cd changed drastically as compared to retention of other metals, this result is consistent with the results of the previous studies in which it was shown that binding capacity of PEI and Cd is maximum at pH 6.5 [6].

It can be concluded that, by adjusting the precipitation parameters (pH, time, precipitation agents), together with the parameters of PEUF experiments (pH and loading (metal/polymer ratio)) selective separation of heavy metals from industrial wastes can be achieved. Process sequencing can be suggested for the removal and recovery of heavy metals, especially for Cd and Cu, from electroplating industry's waste effluents.

REFERENCES

1. Electronic Code of Federal Regulations (e-CFR) TITLE 40--Protection of Environment Chapter I-EPA: Subchapter N-Effluent Guidelines and Standards
2. Peters, R.W.; Ku, Y. AIChE Symposium Series: Separation of heavy metals and other trace contaminants.1985, p.9.
3. Mulder M., Basic Principles of Membrane Technology, Kluwer Academic Publishers, 1st ed.,1991, Netherlands.
4. Mundkur S.D., Watters J.C., Polyelectrolyte Enhanced Ultrafiltration of Copper from a Waste Stream, Separation Science and Technology, 28,pp.1157,1993.
5. Geckeler K.E., Volchek K., "Removal of Hazardous Substances from Water Using Ultrafiltration in Conjuction with Soluble Polymers", Critical Review, Environmental Science and Technology, 30(3), 1996.
6. Müslehiddinoğlu J.,Uludağ Y., Önder H.Ö., Yılmaz L.,"Effect of Operating Parameters on Selective Separation of Heavy Metals from Binary Mixtures via Polymer Enhanced Ultrafiltration", Journal of Membrane Science, 140,p.251-266., 1998.

137c New Approach for Addressing Environmental Effects through Multi Objective Multi Refinery Optimization

Ahmed S. Khogeer, Research & Development Center, Saudi Aramco, PO Box 5836, Saudi Aramco, Dhahran, 31311, Saudi Arabia and M. Nazmul Karim, Colorado State University, 100 Glover Building, Fort Collins, CO 80523.

Petroleum refining is a capital-intensive business. With stringent environmental regulations on the processing industry and declining refining margins, higher pressures are exerted on refiners to optimize performance and find the best combination of feed and processes to produce salable products that meet stricter product specifications, while at the same time meeting refinery supply commitments and of course making profit. Usually it is hard quantify environmental effects on big refining projects either during the design phase or the operating phase. Methods such as sustainability may not be easy to understand by corporate executives and decision maker in many countries. This work concentrate on finding ways to quantify these effects and the possible way for product specification upgrade and embedding this in a multi refinery optimization model. This is done through multi objective optimization. Most refinery-wide optimization methods do not cover multiple objectives such as minimizing environmental impact, This work starts by carrying out a refinery-wide, single objective optimization, and then it moves to multi objective-single optimization. The last step is multi objective-multi refinery optimization, the objectives of which are analysis of the effects of economic, environmental, product spec, Simulation runs were carried out using both MATLAB and ASPEN PIMS utilizing nonlinear techniques to solve the optimization problem. The results addressed the need to debottleneck some refineries or transportation media in order to meet the demand for essential products

137d Effective Field Characterization of a Groundwater Tce in Fractured Basalt

Sheri L. Moore[1], Catherine M. Martin[1], and Richard Garrison[2]. (1) Environmental Engineering & Technology Section, US Army Corps of Engineers, 4735 East Marginal Way South, Seattle, WA 98134, (2) Geology, US Army Corps of Engineers, 4735 East Marginal Way South, Seattle, WA 98134

This presentation will provide valuable lessons learned from a recently completed TCE groundwater characterization effort in fractured basalt aquifers and will demonstrate the effectiveness of the Triad approach from project planning, implementation, and evaluation through project completion to regulatory acceptance. The presentation will assess data management and evaluation, team roles, project bottlenecks, data visualization, data gaps and uncertainty management, and regulatory acceptance.

In 2004 and 2005, U.S. Army Corps of Engineers, Seattle District, conducted a field investigation that combined multiple investigation and analytical methods to further characterize low-level TCE plumes in fractured basalt aquifers at a Superfund site (former Larson Air Force Base in Moses Lake, Washington). The field plan was developed using systematic planning to create a dynamically-structured strategy for the field work to progress in a way to best meet project objectives based on an evolving conceptual site model. The plan was designed to allow the project team to make real-time decisions during field work to complete the remedial investigation phase of work and provide remaining necessary information for a feasibility study. The dynamic approach employed by the project team was dependent upon a detailed management plan (describing investigation methodologies, rationales, and sequencing), multi-discipline field and office teams with a mix of experience and expertise, partnering with regulators, traditional and innovative field investigation methods (drilling: air rotary, sonic, enhanced CPT; sampling: in-situ drilling samples, passive diffusion bag samplers, low-flow sampling), fixed and field analytical techniques (quick-turn fixed laboratory and on-site Direct Sensing Ion-Trap Mass Spectrometry), and progressive electronic data management (SEDD deliverables, web-based data tables and interactive web formats for coordination between field team, contractors, and in-house USACE and EPA teams).

137e Triad Case Studies Lessons Learned from Implementation at Diverse Sites

Sheri L. Moore, Kira Lynch, and Kymberly Takasaki. Environmental Engineering & Technology Section, US Army Corps of Engineers, 4735 East Marginal Way South, Seattle, WA 98134

This presentation will highlight case studies in which the Triad approach was used to reduce costs and time spent in reaching project goals. The case studies represent multiple sites with diverse characteristics where each site benefited because of systematic planning, dynamic field decision strategies, and real-time measurement technologies. Case study sites to be discussed may include: contaminated sediments; brownfields; military range contamination; DNAPL; and widely dispersed, low-level contaminant plumes.

Approaches that maximize real-time analysis, rapid sampling, and dynamic decision-making lead to significant savings in both time and money spent on site cleanup. Investment in a thorough, up-front systematic planning process ensures that site data not only result in a standard consultant's report, but also advance cleanup and land re-use goals with the necessary level of confidence in supporting decisions. This approach – integrating the use of systematic planning, dynamic strategies, and real-time measurement technologies – is referred to as the "Triad." It draws together the lessons learned by practitioners over the past 25 years of site cleanup practice with the technological and knowledge advances that have occurred into a comprehensive framework. The Triad focuses on identifying and addressing all sources of uncertainty at sites and using a thorough conceptual site model to drive data collection. At the growing number of sites where Triad has been applied, a consistent theme of cost savings, shortened project timeframes, and improved decision-making is emerging.

179a Performance Based Management Concepts and Implementation

Javier Santillan, Technical Directorate, Environmental Science Division, HQ Air Force Center for Environmental Excellence, 3300 Sidney Brooks Rd, Brooks City-Base, TX 78235-5112

The AFCEE Technology Transfer Outreach Office developed Restoration PBM Draft Guidance that was successfully tested at DoD sites. PBM restoration incorporates, in its framework, well recognized and accepted business practices and value engineering techniques such as streamlined or accelerated site characterization (i.e, Triad), Remedial Process Optimization (RPO), development of a definitive exit strategy based on attainable Applicable or Relevant and Appropriate Requirements (ARARs), and risk management based on future land use. PBM also promotes performance-based restoration, including MMRP and low-level radiological waste, by procuring services through performance-based contracts.

This course will provide in depth coverage of each component of PBM as well as case studies where the concepts were applied.

179b Intergovernmental Data Quality Framework, Mir, Ufp

Edward Brown, TDE, Air Force Center for Environmental Excellence, 3300 Sidney Brooks, Brooks City-Base, TX 78235
QAPS and associated efforts to maintain data quality have become more and more complicated over the past several years. Quite often, the documentation has not kept up with new analytical methodologies or approaches.

This short course will provide participants with an understanding of two new policies to address analytical quality: Uniform Federal Policy (UFP) for Implementing Environmental Quality Systems and the Methods Innovation Rule (MIR).

Headquarters U.S. Air Force

Integrity - Service - Excellence

Implementation of Remedial Process Optimization within the Air Force

Mr. John Gillespie
HQ AFCEE/TDE

U.S. AIR FORCE

U.S. AIR FORCE

What is Remedial Process Optimization?

- Remedial Process Optimization (RPO)
  - A systematic approach for evaluating existing cleanup and monitoring systems with the goal of improving their effectiveness and reducing overall site cleanup costs without increasing risks
  - A component of the overall Air Force Performance-Based Management of Cleanup Program
- RPO applies to
  - All cleanup activities, both interim and final
  - From the Feasibility Study (FS) through SC phases
  - Operating Remedial Action (RA) systems
  - Sites undergoing long-term monitoring (LTM) activities

Integrity - Service - Excellence

2

U.S. AIR FORCE

RPO - Component of PBM

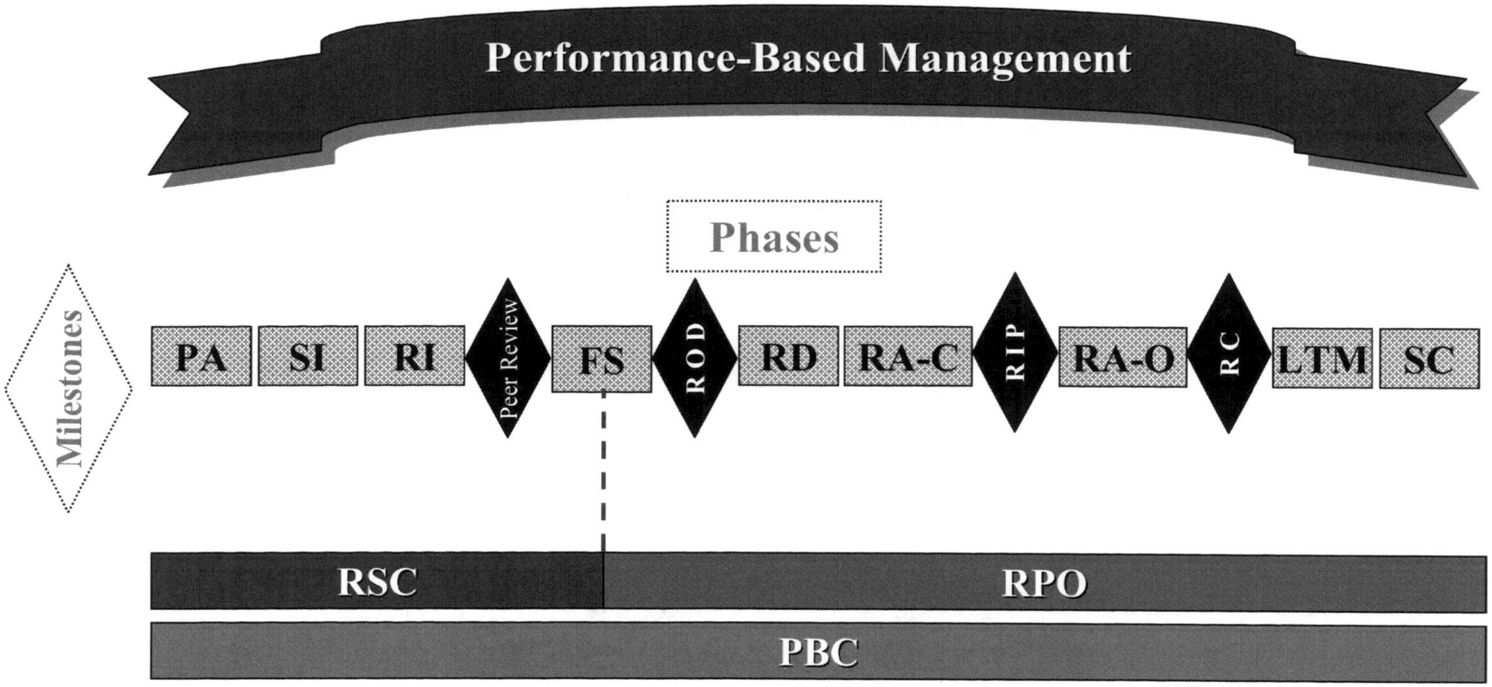

RPO is effective…from the FS to SC

U.S. AIR FORCE

Where does RPO fit into the Clean-up Scheme?

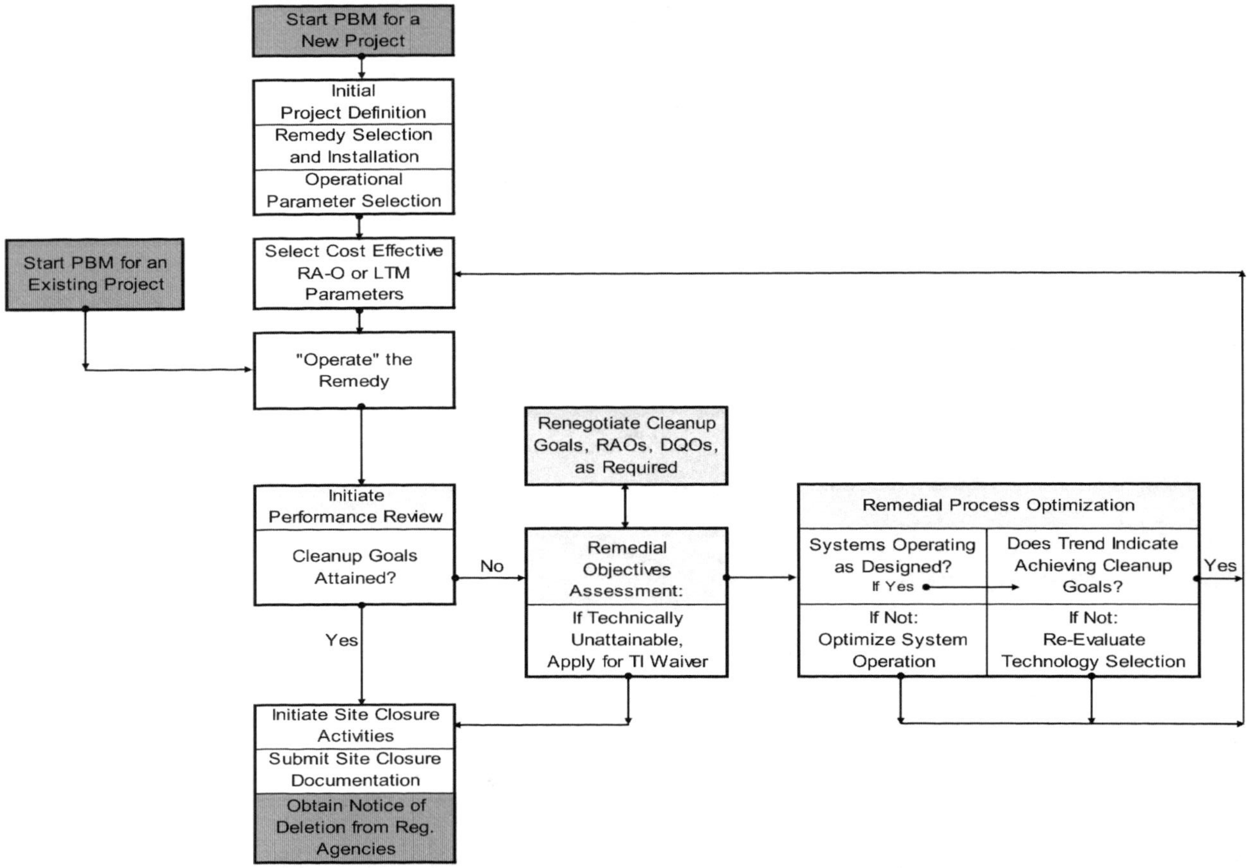

Integrity - Service - Excellence

4

U.S. AIR FORCE

Why RPO?

- The Air Force strategy is to move away from active remediation systems

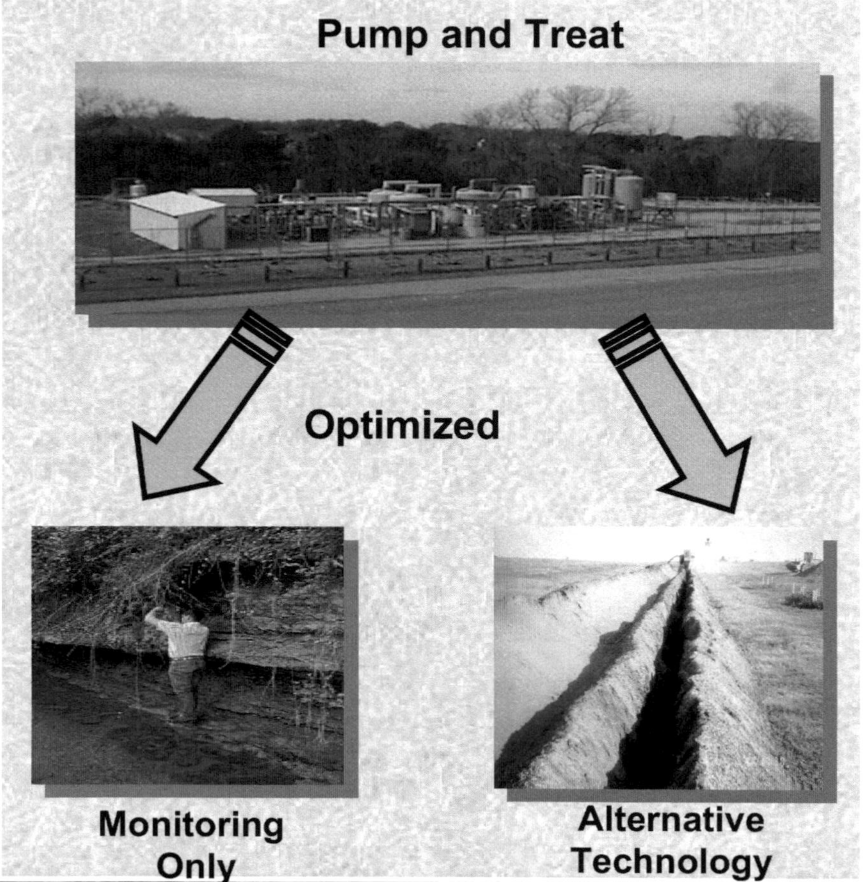

U.S. AIR FORCE

RPO Program Goals

- Ensure cleanup actions achieve protection of human health and the environment, and are working towards accomplishing the cleanup objectives, goals, and levels outlined in a decision document
- Continually re-evaluate site cleanup strategies and processes, considering recent changes in legal requirements, site environmental factors, and technological options
- Track and report cleanup progress
- Optimize in-place cleanup systems and LTM to minimize operation and maintenance (O&M) cost
- Accelerate site transfer or closure, achieving site cleanup levels more rapidly and efficiently

U.S. AIR FORCE

When do we start doing RPO?

- RPO begins in the Feasibility Study/Corrective Measures Study Phase
  - Optimization principles shall be applied during the planning, design, and construction of cleanup actions
  - When selecting a cleanup method, lifecycle cost of the system should be a major consideration, not just the initial capital costs
  - The FS/CMS report shall identify conditions for SC and lifecycle cost analysis through SC

U.S. AIR FORCE

How Do We Collect Pertinent Environmental Information for the Optimization Program?

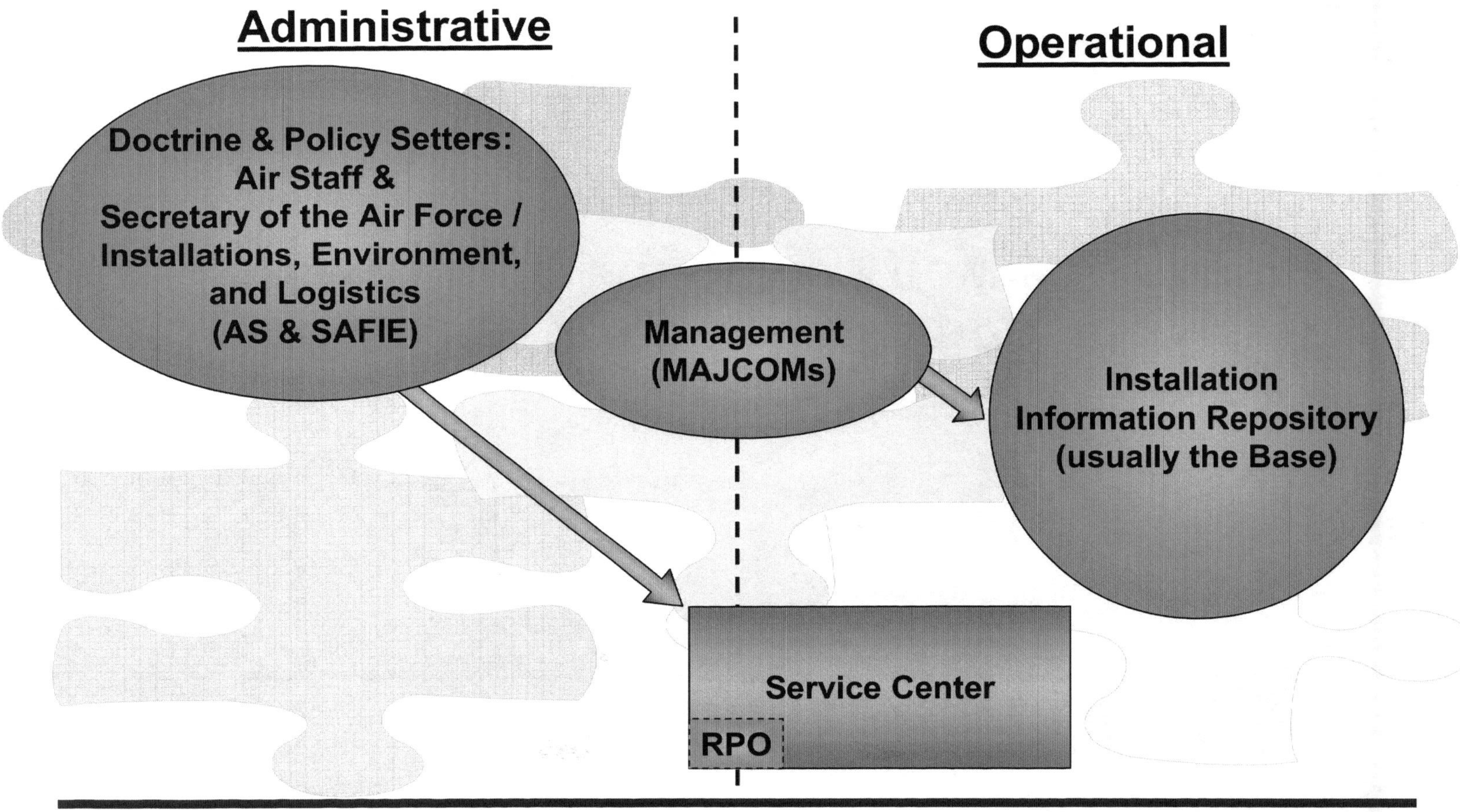

Integrity - Service - Excellence

U.S. AIR FORCE

Air Force Optimization Challenge

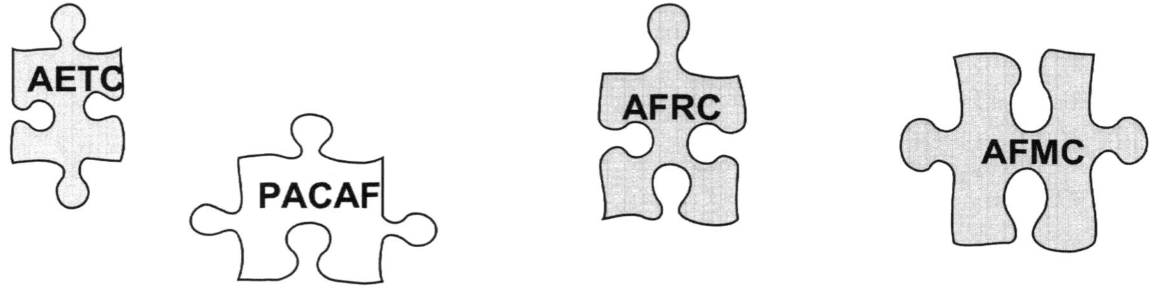

How do we integrate optimization into the Air Force without adequate knowledge base?

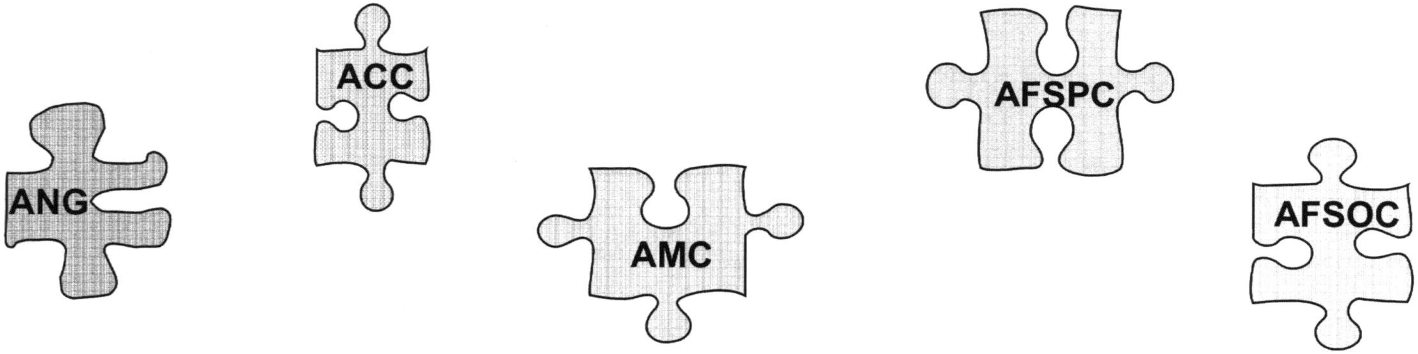

U.S. AIR FORCE

Integrating RPO

- Where is the center of gravity?
- Unlike the EPA…
  - Air Force is set up by function vs. geographically
  - MAJCOM vs. EPA Region

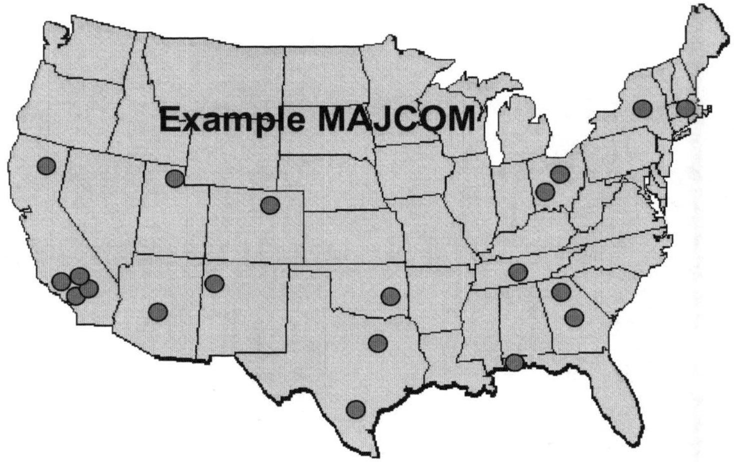

Great for warfighting, but a challenge for addressing environmental issues

U.S. AIR FORCE

RPO Business Model to Collect Information

- Air Staff created RPO Outreach Office at AFCEE to assist MAJCOMs with their optimization programs
  - Outreach Office Objectives:
    - Provide a standardized approach across Air Force
    - Help MAJCOMs build an inventory, create an optimization baseline, track trends, & integrate optimization decision making process
    - Used EPA web-based questionnaire as building block to develop RPO Inventory & Performance Software (RIPS)
    - Emphasize performance & results

U.S. AIR FORCE

Vital Optimization Questions...

- **Technology?**
  - What technology is implemented?
  - What is the average life-cycle?
  - What is the legal driver for the remediation system?
  - What is the primary remedial objective as currently operating versus original design?
  - Are the remedial objectives being achieved?
  - What are the primary COCs?
- **Cost?**
  - What is the system O&M cost?
  - What is the average cost for each technology?
  - What is the projected life-cycle cost?

Three-Phased Approach

- Phase I – Inventory and Review
- Phase II – Evaluations
- Phase III – Implementation of Recommendations

U.S. AIR FORCE

Remedial Process Optimization Sequence

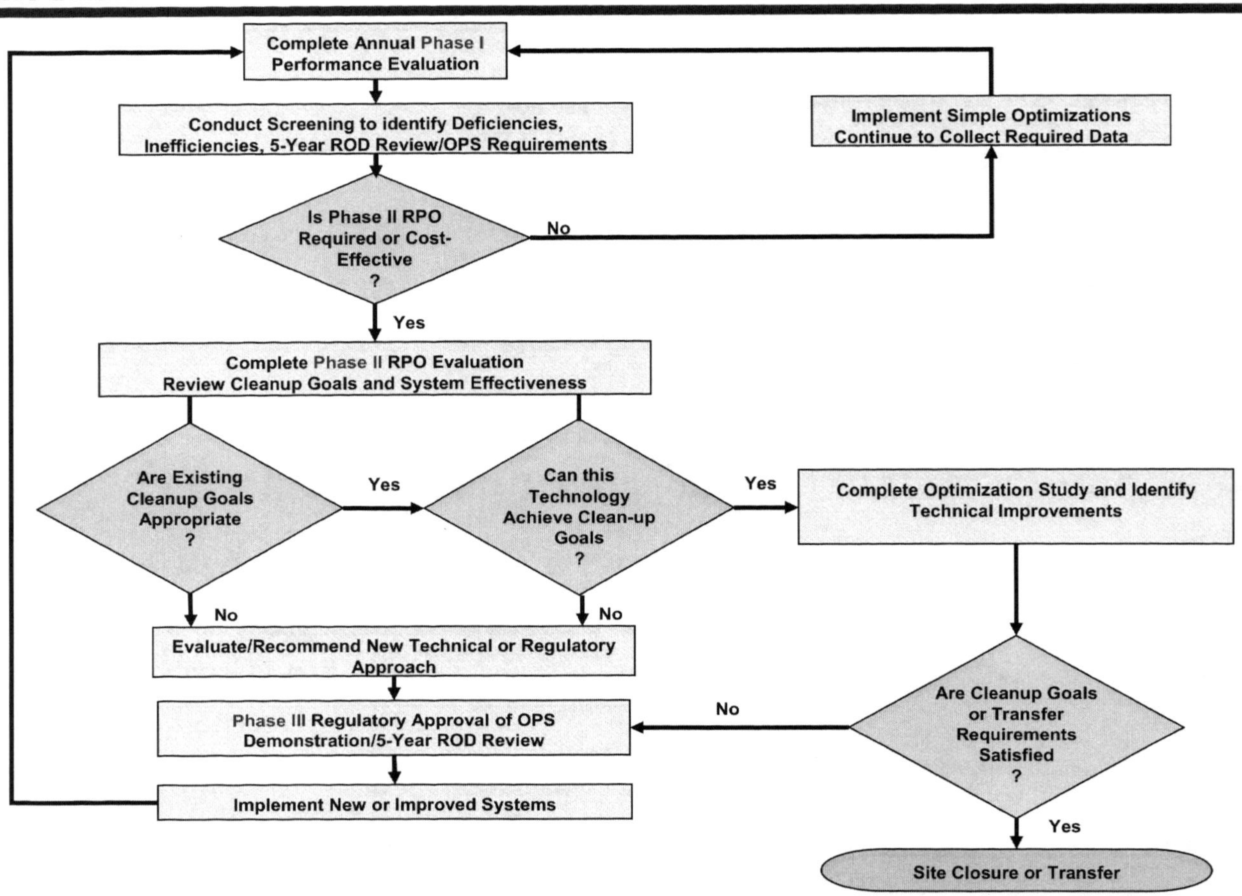

Integrity - Service - Excellence

14

U.S. AIR FORCE

Phase I - Inventory and Review

- Inventory and review of system performance shall be accomplished annually for each system and site under active remediation or LTM
- Ideally, data collection for Phase I inventory and review will be accomplished throughout the year
- Cost accounting of the O&M of systems/monitoring and details of system performance shall be tracked and recorded
- The minimum requirement for an annual Phase I inventory and review is the completion of an RPO Inventory & Performance Software (RIPS) inventory and assessment

U.S. AIR FORCE

Phase I Software

- RPO Inventory & Performance Software (RIPS) is the cornerstone of the Air Force RPO
- Web-based database
  - RIPS provides an AF-wide, comprehensive inventory of remediation systems and LTM programs
- Knowledge management tool
  - Assists in prioritization of systems in terms of optimization potential
- Performance tracking capability in FY06
  - Allow for benchmarking technologies
- RIPS is updated annually
  - Data cut-off is end of FY, complete entry by end of CY

RIPS is a tool...

U.S. AIR FORCE

RIPS Screenshot - *Main Menu*

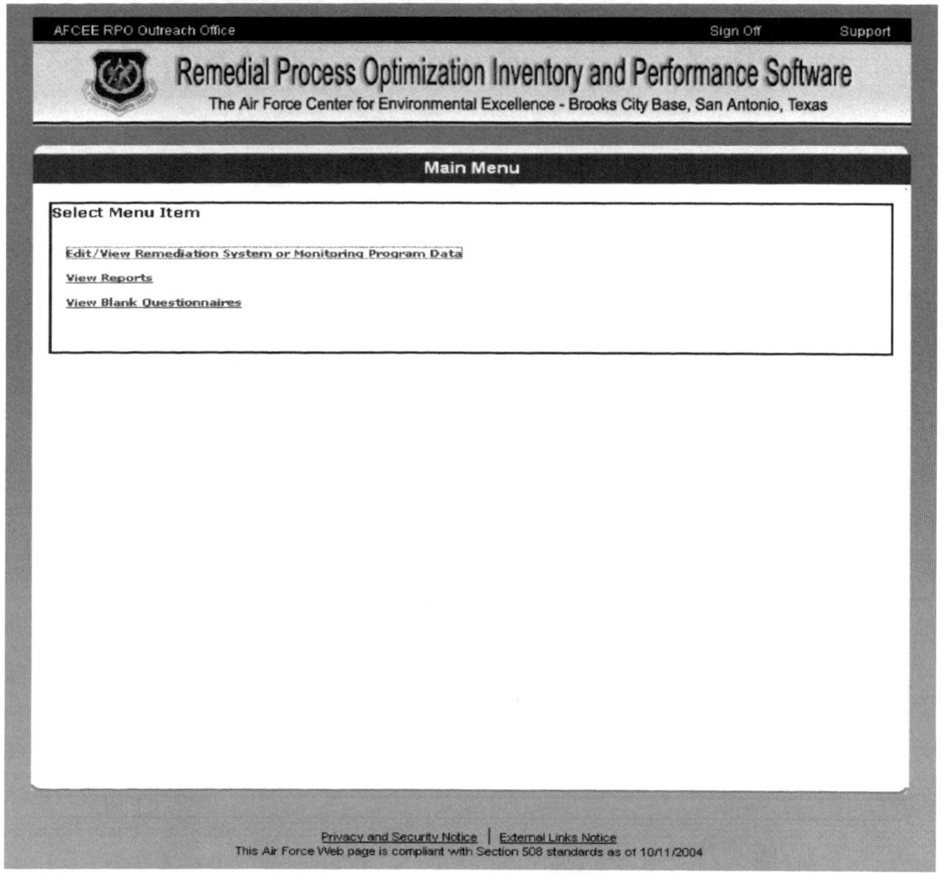

Integrity - Service - Excellence

17

U.S. AIR FORCE

RIPS Screenshot – System/Program Selection

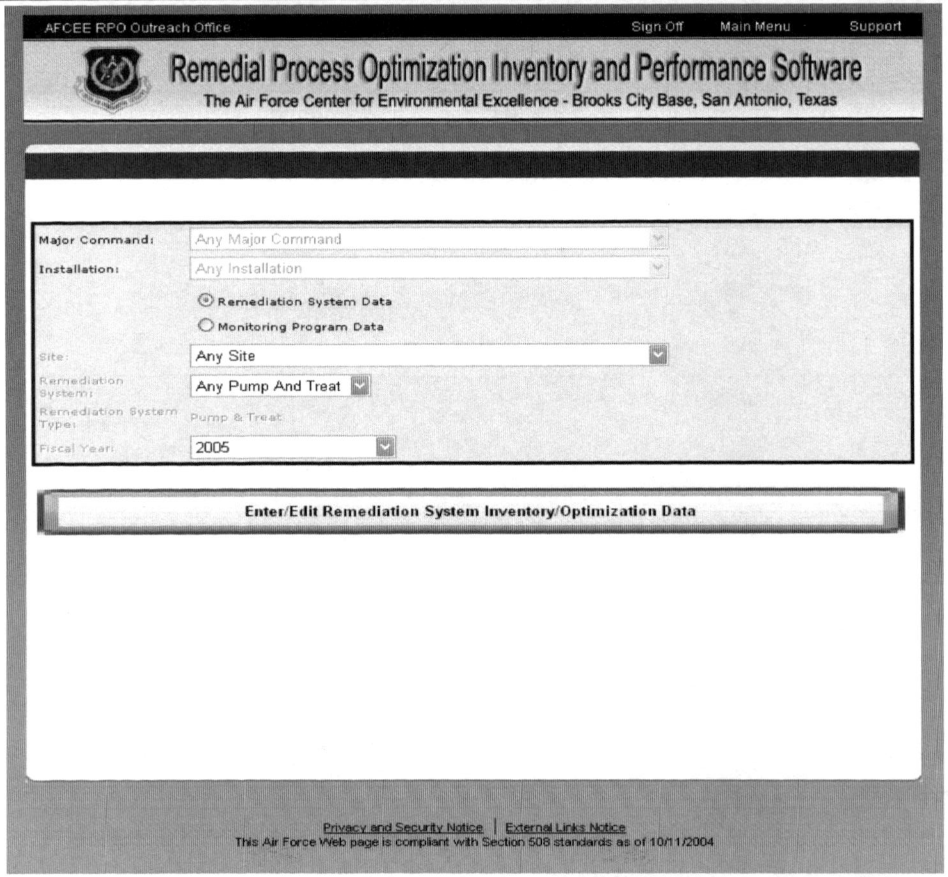

Integrity - Service - Excellence

RIPS Screenshot – Blank Questionnaire

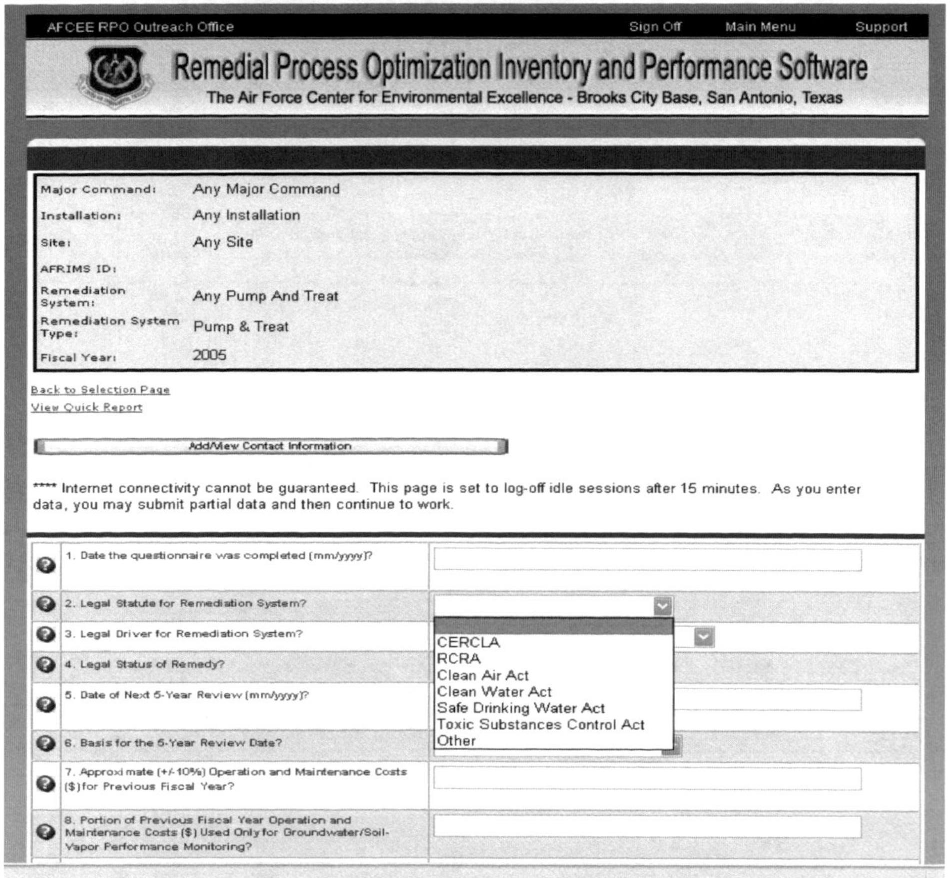

Integrity - Service - Excellence

19

U.S. AIR FORCE

RIPS Screenshot – Complete Questionnaire

AFCEE RPO Outreach Office — Sign Off — Main Menu — Support

Remedial Process Optimization Inventory and Performance Software

The Air Force Center for Environmental Excellence - Brooks City Base, San Antonio, Texas

Major Command:	[redacted]
Installation:	[redacted]
Site:	[redacted]
AFRIMS ID:	
Remediation System:	East Parking Lot GW Treatment System
Remediation System Type:	Pump & Treat
Fiscal Year:	2005

Back to Selection Page — View Quick Report — Add/View Contact Information — Edit/Add Performance Tracking Tool Information

**** Internet connectivity cannot be guaranteed. This page is set to log-off idle sessions after 15 minutes. As you enter data, you may submit partial data and then continue to work.

1. Date the questionnaire was completed (mm/yyyy)?	10/2005
2. Legal Statute for Remediation System?	CERCLA
3. Legal Driver for Remediation System?	Record of Decision (ROD)
4. Legal Status of Remedy?	Final
5. Date of Next 5-Year Review (mm/yyyy)?	02/2008
6. Basis for the 5-Year Review Date?	RA-O Start Date
7. Approximate (+/-10%) Operation and Maintenance Costs ($) for Previous Fiscal Year?	$210,000.
8. Portion of Previous Fiscal Year Operation and Maintenance Costs ($) Used Only for Groundwater/Soil-Vapor Performance Monitoring?	$120,000.00
9. Approximate (+/-10%) Operation and Maintenance Costs ($) for Current Fiscal Year?	$200,000.00
10. Portion of Current Fiscal Year Operation and Maintenance Costs ($) Used Only for Groundwater/Soil-Vapor Performance Monitoring?	$120,790.00

12. Expected Completion Date for Remedial Action Based on Remedial Design (mm/yyyy)?	09/2015
13. Expected Completion Date for Remedial Action as Currently Operating (mm/yyyy)?	09/2035
14. Primary Remedial Objective Based on Remedial Design?	Restoration
15. Primary Remedial Objective as Currently Operating?	Both High-Concentration Area Reduction and Containment
16. Are the Remedial Objectives from the Remedial Design Being Achieved at Present?	Yes
17. Are the Remedial Objectives as Currently Operating Being Achieved at Present?	Yes
18. Are the Remedial Objectives Based on the Remedial Design Reasonable?	Yes
19. Are the Remedial Objectives as Currently Operating Reasonable?	Yes
20. Has an Independent/Third-Party Evaluation of System Effectiveness and Efficiency Been Completed?	No
21. What is the Consequence/Risk of System Failure?	High
22. Expected Degree of Social/Political/Regulatory Difficulty in Implementing Minor Changes to the Remediation System?	Little difficulty
23. Expected Degree of Social/Political/Regulatory Difficulty in Implementing Major Changes to the Remediation System?	Moderate difficulty
24. Is there a Documented Conceptual Site Model?	Yes
25. When was the Last Documented Update of the CSM (mm/yyyy)?	11/2002
26. Is LNAPL currently present?	No
27. Is DNAPL currently present?	Yes
28. Highlight the Primary COC(s) for the Remediation System (to make multiple selections, use the mouse while depressing the [Ctrl] key):	Chlorinated Hydrocarbons (e.g. TCE, TCA, DCE, vinyl chloride)
29. Is the System Successfully Controlling Plume Migration?	Yes
30. Number of Groundwater Extraction Wells (Total)?	52
31. Number of Groundwater Extraction Wells (In Use)?	52
32. Number of Groundwater Injection Wells (Total)?	0
33. Number of Groundwater Injection Wells (In Use)?	0
34. Number of Horizontal Groundwater Extraction Wells (Total)?	1
35. Number of Horizontal Groundwater Extraction Wells (In Use)?	1
36. Number of Groundwater Extraction\Interceptor Trenches (Total)?	0
37. Number of Groundwater Extraction\Interceptor Trenches (In Use)?	0

38. Number of Groundwater Infiltration Galleries (Total)?	0
39. Number of Groundwater Infiltration Galleries (In Use)?	0
40. Groundwater Extraction Rate (gallons/minute)?	100
41. Groundwater Injection Rate (gallons/minute)?	0
42. Estimated Average Operational Efficiency (percent)?	77
43. Number of Groundwater Wells Regularly Monitored?	5
44. Frequency of Groundwater Monitoring (events/year)?	2
45. Highlight One or More of the Following Aboveground Treatment Processes (to make multiple selections, use the mouse while depressing the [Ctrl] key):	Air Stripping
46. Please provide any general comments, or a brief explanation for any of the above questions that were not answered or were answered "Other".	$$ do not include labor, management fee of $83K, DNAPL suspected at one time, but concentrations have decreased.

Integrity - Service - Excellence

U.S. AIR FORCE

RIPS Report

Remediation System Operation and Maintenance Costs by System Type - 2005
(Sorted by Cost Reduction)

	Major Command	Installation	Site	Remediation System	Previous Fiscal Year O&M Costs	Current Fiscal Year O&M Costs	Cost Reduction [1]
				Dual-Phase (Groundwater & Soil Gas) Extraction			
1	Air Force Materiel Command	[redacted]	LF03	Land Fill 3 Remediation System	$40,000	$38,000	$2,000
2	Air Force Materiel Command	[redacted]	Building 181 AOC	Building 181 SVE Remediation System	$55,000	$53,135	$1,865
				Totals:	**$95,000.00**	**$91,135.00**	**$3,865.00**
				Pump & Treat			
1	Air Force Materiel Command	[redacted]	FSA#1	Fuel Saturation Area #1 Remediation System	$80,000	$68,933	$11,067
2	Air Force Materiel Command	[redacted]	OT22	East Parking Lot GW Treatment System	$210,000	$200,000	$10,000
3	Air Force Materiel Command	[redacted]	Carswell AFB LF4/5	LF 4/5 Pump and Treat Treatment System	$0	$0	$0
				Totals:	**$290,000.00**	**$268,933.00**	**$21,067.00**
				Thermal Technology (e.g., electrical resistance heating, steam enhanced extraction)			
1	Air Force Materiel Command	[redacted]	Building 181 AOC	Building 181 Thermal Technology Remediation System	$0	$0	$0
				Totals:	**$0.00**	**$0.00**	**$0.00**
				Wall/Barrier System (e.g., slurry wall, permeable reactive barrier)			
1	Air Force Materiel Command	[redacted]	Carswell AFB LF4/5	LF 4/5 Permeable Wall Remedation System	$40,000	$20,000	$20,000
				Totals:	**$40,000.00**	**$20,000.00**	**$20,000.00**
				GrandTotals:	**$425,000.00**	**$380,068.00**	**$44,932.00**

(1) Cost reduction = Previous Fiscal Year O&M costs - Current Fiscal Year O&M costs.

Installations Selected for

Air Force Plant 4

Integrity - Service - Excellence

U.S. AIR FORCE

RIPS Data

- **The Air Force was able to identify potential TI waiver candidates using data submitted into RIPS**

System Type	Expected Completion Date as Currently Operating:	$ Based on Years Remaining	Legal Driver for Remediation System:	Legal Statute for Remediation System:	O&M Costs for Current Fiscal Year:
Wall/Barrier System (e.g., slurry wall, permeable reactive barrier)			Interim Record of Decision (IROD)	CERCLA	$388,000
Pump & Treat	12/2055	$143,860,470	Interim Record of Decision (IROD)	CERCLA	$2,880,050
Pump & Treat	07/2047	$76,673,906	Record of Decision (ROD)	CERCLA	$1,846,406
Pump & Treat	08/2044	$47,093,902	Interim Record of Decision (IROD)	CERCLA	$1,219,703
Pump & Treat	09/2045	$46,047,947	Interim Record of Decision (IROD)	CERCLA	$1,160,018
Pump & Treat	01/2033	$27,021,918	Record of Decision (ROD)	CERCLA	$1,000,000
Pump & Treat	07/2020	$26,779,855	Interim Record of Decision (IROD)	CERCLA	$1,845,666
Pump & Treat	06/2130	$22,409,753	Interim Record of Decision (IROD)	CERCLA	$180,000
Pump & Treat	07/2029	$16,543,321	Record of Decision (ROD)	CERCLA	$703,520
Pump & Treat	01/2034	$12,133,490	Voluntary Action	CERCLA	$433,000
Pump & Treat	01/2037	$11,044,778	Voluntary Action	CERCLA	$356,000
Pump & Treat	09/2020	$6,896,407	Interim Record of Decision (IROD)	CERCLA	$469,800
Pump & Treat	06/2049	$6,741,562	Other	CERCLA	$155,169
Pump & Treat	06/2023	$6,099,589	Record of Decision (ROD)	CERCLA	$350,000
Pump & Treat	05/2017	$5,669,863	Record of Decision (ROD)	CERCLA	$500,000
Pump & Treat	12/2045	$3,303,468	Interim Record of Decision (IROD)	CERCLA	$82,700
Monitored Natural Attenuation	11/2055	$10,123,304	Interim Record of Decision (IROD)	CERCLA	$203,000
Dual-Phase (Groundwater & Soil Gas) Extraction	06/2199	$40,644,493	Interim Record of Decision (IROD)	CERCLA	$210,000
Dual-Phase (Groundwater & Soil Gas) Extraction	06/2130	$12,449,863	Interim Record of Decision (IROD)	CERCLA	$100,000
Dual-Phase (Groundwater & Soil Gas) Extraction	01/2016	$5,994,923	Voluntary Action	CERCLA	$599,000
top 20					
completion date					
cost					
IROD					
$ remaining					

Integrity - Service - Excellence

22

U.S. AIR FORCE

Total FY05 Cost

All MAJCOMs

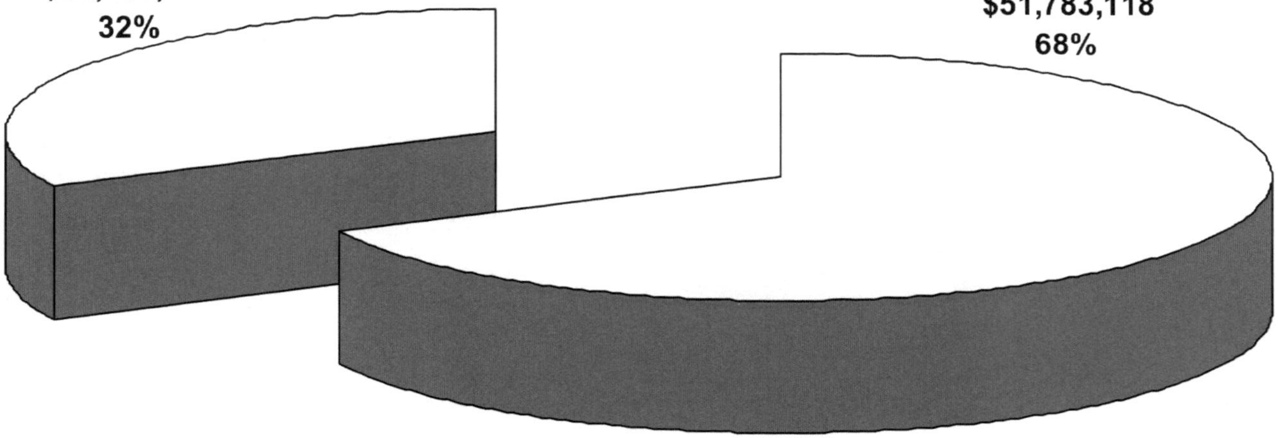

Total FY05 Cost = $76,552,355

*As of 28 Feb 06

Integrity - Service - Excellence

U.S. AIR FORCE

Total FY05 Number of Systems by Technology

System Type – All MAJCOMs

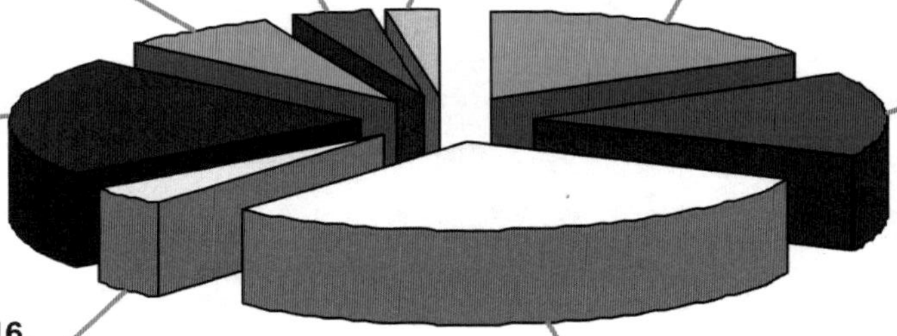

*As of 28 Feb 06

Integrity - Service - Excellence

U.S. AIR FORCE

Total FY05 System Cost by Technology

System Cost – All MAJCOMs

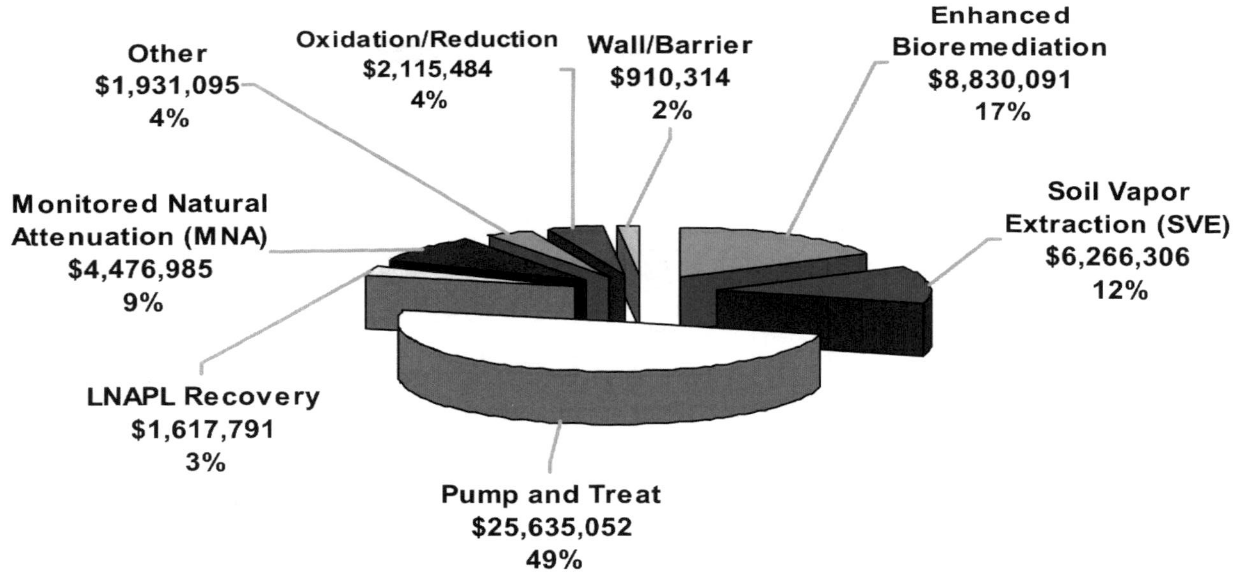

*As of 28 Feb 06

U.S. AIR FORCE

Average System Life-Span by Type

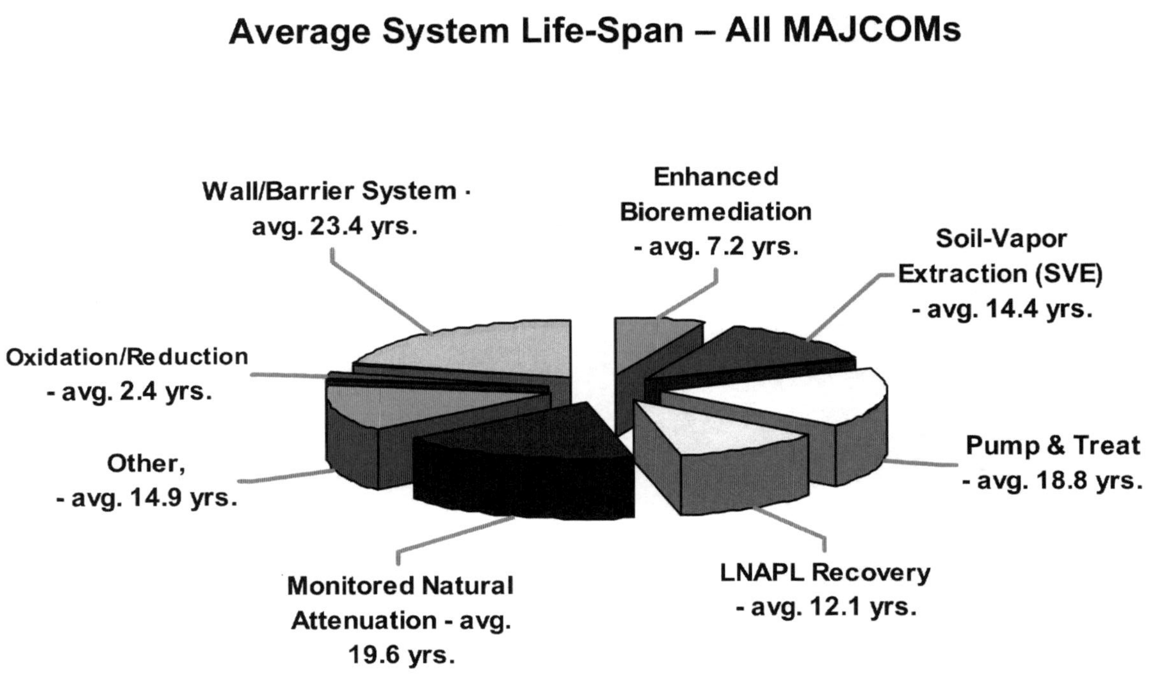

Integrity - Service - Excellence

26

U.S. AIR FORCE

Phase II – LTM Evaluations

- LTM systems
  - Generally, offers the highest payback and least regulatory resistance – particularly on "un-optimized" systems
  - Monitoring and Remediation Optimization Software (MAROS)
  - Geostatistical Temporal and Spatial (GTS) optimization software
  - Professional Judgment of Environmental Scientists and Engineers

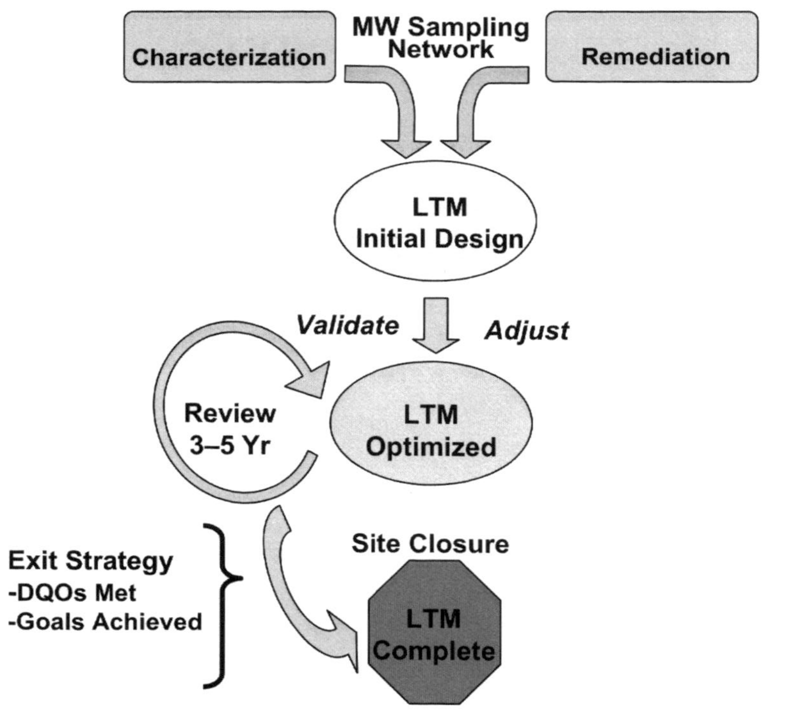

U.S. AIR FORCE

Phase II LTM - Optimization Identifies

Sampling Redundancy & Essential Data

"Nice to have"
All Data
Samples = 240

"Essential"
90% Reduction
Samples = 27

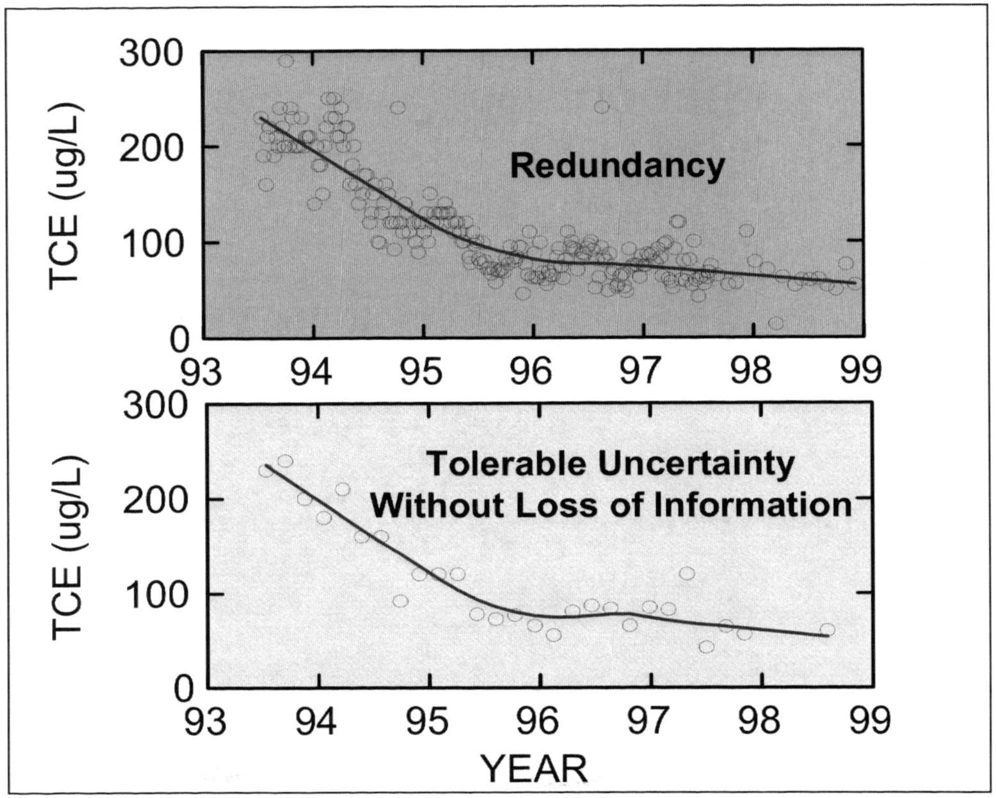

Integrity - Service - Excellence

28

U.S. AIR FORCE

Phase II LTM - Resource Requirements

- Data Driven, need electronic data (ERPIMS)
- Conceptual site model (CSM)
- Funding \$10-35K per site
- Entire installation can be > \$200K
- Typically < 15% annual O&M budget

U.S. AIR FORCE

Phase II LTM - Tool Selection

- GTS (Geostatistical Temporal-Spatial) algorithm
  - Minimum no. wells > 30
  - Multiple sites and OUs, or individual sites
  - Spatial/temporal algorithm: locally weighted quadratic regression (LWQR)
  - Need a geostatistician
- MAROS
  - Site modeled as a distinct plume
    - Source area known
    - Plume delineation & stability analysis
    - Well categories respective of plume
  - Handles large sites or networks < 40 wells
  - Spatial/temporal algorithms: Delaunay method and modified CES, respectively

U.S. AIR FORCE

Phase II Remediation System – Evaluations

- Cleanup systems
  - Evaluation can recommend minor changes or system shutdown and replacement with alternative treatment
  - Optimization must be implemented within the confines of existing decision document
  - Engage and include regulatory and public stakeholders early and often

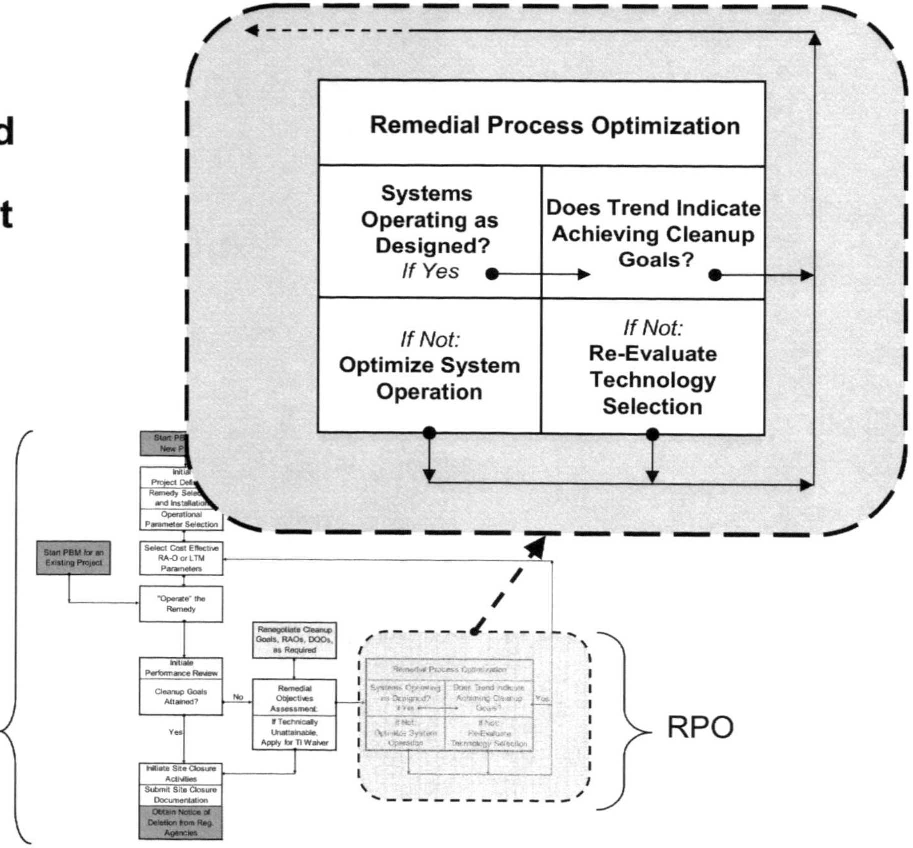

U.S. AIR FORCE

Effectiveness and Efficiency

- First – verify effectiveness
  - Is the system protective?
  - Will the system achieve site closure?

- Then – analyze efficiency
  - Is performance in-line with cleanup objectives, goals, and levels defined within the performance-based decision document?
  - Is system optimization necessary?
    - Accelerate cleanup
    - Reduce cost

Confirm effectiveness before evaluating efficiency...

Integrity - Service - Excellence

U.S. AIR FORCE

Phase II – Evaluations

- Prioritize systems for Phase II evaluations and determine the level of effort
- Balance consideration of protectiveness, cost-to-complete, and schedule
- Systems not protective of human health and the environment as outlined in the decision document should receive the highest priority
- Sites with upcoming five-year reviews receive priority
  - RPO evaluations shall be conducted such that, necessary data and analysis will be available to support meaningful discussions and negotiations during a five-year review

Integrity - Service - Excellence

U.S. AIR FORCE

Phase II – Evaluations

- Recommended a team consisting of a combination of contractor support, MAJCOM representative(s), and RPMs from other installations be utilized
- Without incentives, contractors currently operation systems may not aggressively pursue optimization, especially if it leads to system shutdown
- It's often difficult for Installation RPMs, with time and effort invested in a remedy, to evaluate it impartially
- Coordination with Staff Judge Advocate and AFCEE Regional Environmental Office is vital
- Incorporating regulators from the Interstate Technology & Regulatory Council (ITRC) is encouraged and may be helpful in gaining cooperation from the regulatory community

U.S. AIR FORCE

Phase II – Conceptual Site Model

- Elements of a Conceptual Site Model
  - Background Information
  - Contaminant Source and Release Information
  - Geologic and Hydrologic Information
  - Contaminant Distribution, Transport, and Fate Parameters
  - Risk Assessment

Background Information

- Location of water supply wells.
- Ground-water classification.
- Nearby wellhead protection areas or sole-source aquifers.
- Locations of potential receptors exposure points.

Contaminant Source and Release Information

- Location, nature, and history of previous contaminant releases or sources.
- Locations and characterizations of continuing releases or sources.
- Locations of subsurface sources (e.g., DNAPLs).
- Flux of contamination from DNAPL.

Risk Assessment

- Current and future receptors.
- Exposure scenario's.
- Completed pathways?
- Exposure concentrations.

Geologic and Hydrologic Information

- Description of regional and site geology.
- Physical properties of subsurface materials (e.g., porosity, bulk density).
- Stratigraphy, including thickness, lateral extent, continuity of units, and presence of depositional features, such as channel deposits, that may provide preferential pathways for, or barriers to, contaminant transport.
- Geologic structures that may form preferential pathways for DNAPL migration or zones of accumulation.
- Depth to ground water.
- Hydraulic gradients (horizontal and vertical).
- Hydraulic properties of subsurface materials (e.g., hydraulic conductivity, storage coefficient, effective porosity) and their directional variability (anisotropy).
- Spatial distribution of soil or bedrock physical/hydraulic properties (degree of heterogeneity).
- Temporal variability in hydrologic conditions
- Groundwater recharge and discharge information.
- Groundwater/surface water interactions.

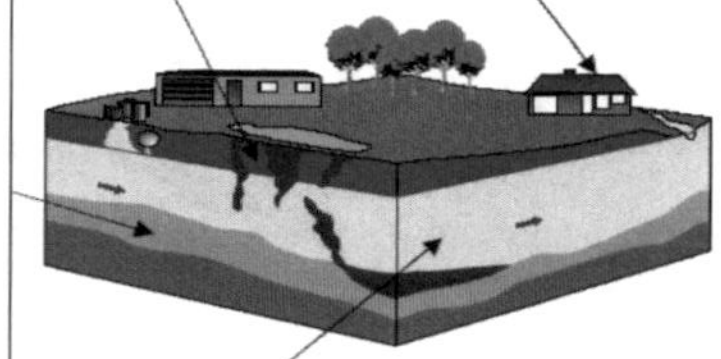

Contaminant Distribution, Transport, and Fate Parameters

- Properties of DNAPLs that affect transport (e.g., composition, effective constituent solubilities, density, viscosity).
- Phase distribution of each contaminant (gaseous, aqueous, sorbed, free-phase DNAPL or residual DNAPL) in the unsaturated and saturated zones.
- Spatial distribution of subsurface contaminants in each phase in the unsaturated and saturated zones.
- Estimates of subsurface contaminant mass.
- Temporal trends in contaminant concentrations in each phase.
- Partitioning coefficients and migration rates.
- Contaminant natural attenuation processes (destructive and non-destructive).

Integrity - Service - Excellence

U.S. AIR FORCE

Phase III - Implementation

- MAJCOMs are responsible for evaluating Phase II RPO reports and determining whether the recommendations shall be implemented
- Changes to systems must be accomplished within the regulatory framework and the signed decision document for the site

U.S. AIR FORCE

Data Mining

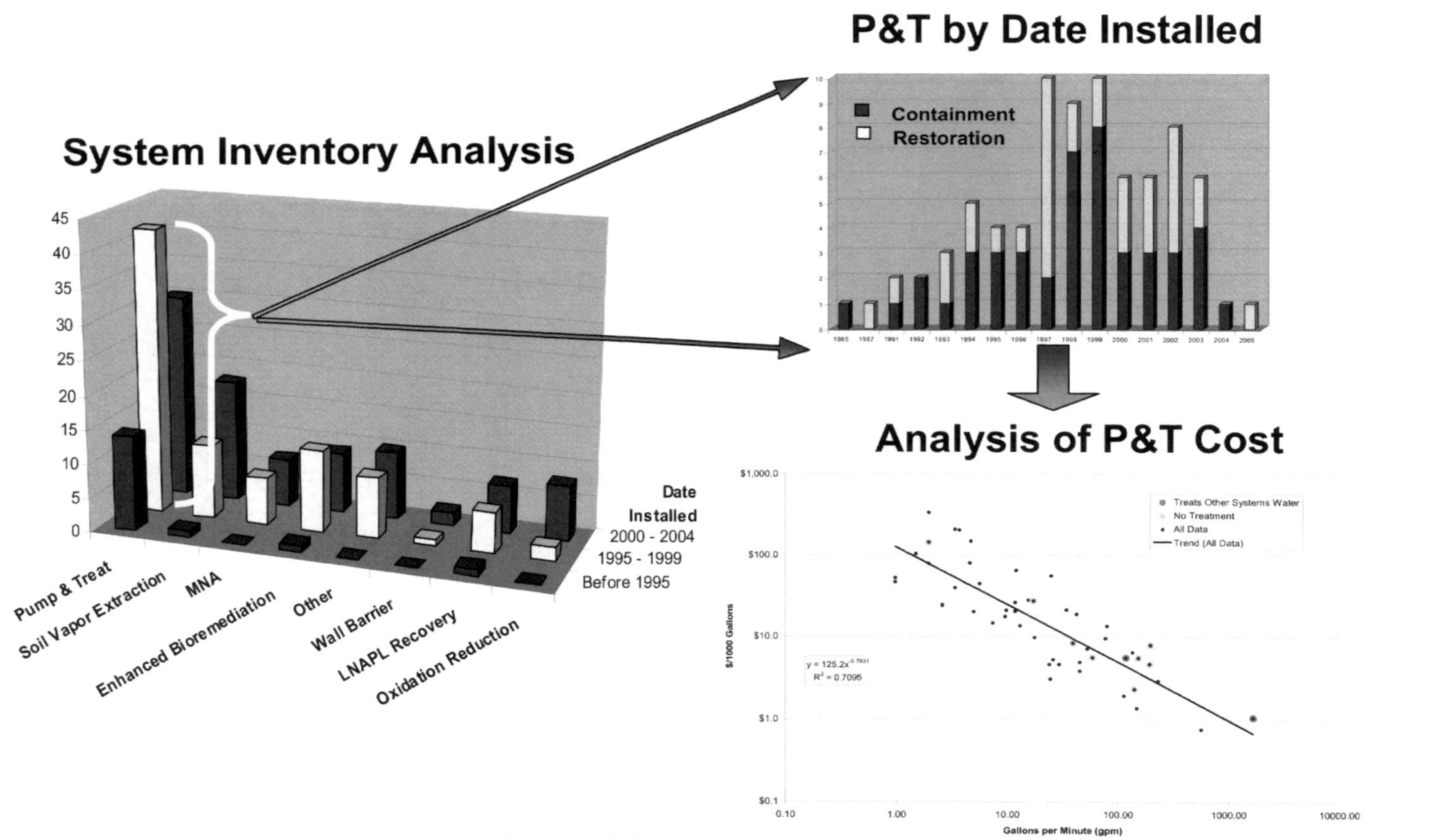

Integrity - Service - Excellence

RPO for Pump and Treat

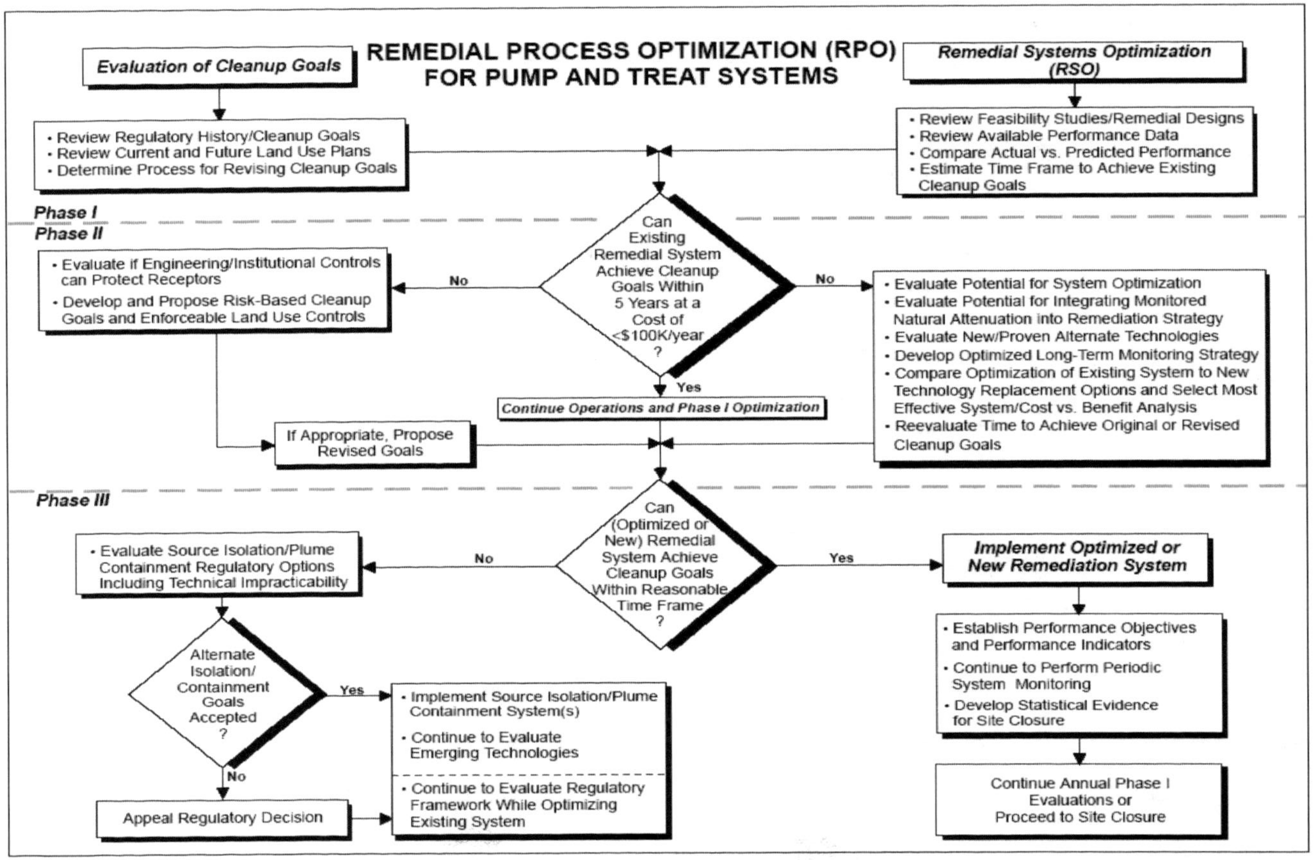

Integrity - Service - Excellence

U.S. AIR FORCE

RIPS Screenshot – *Performance Tracking Tool*

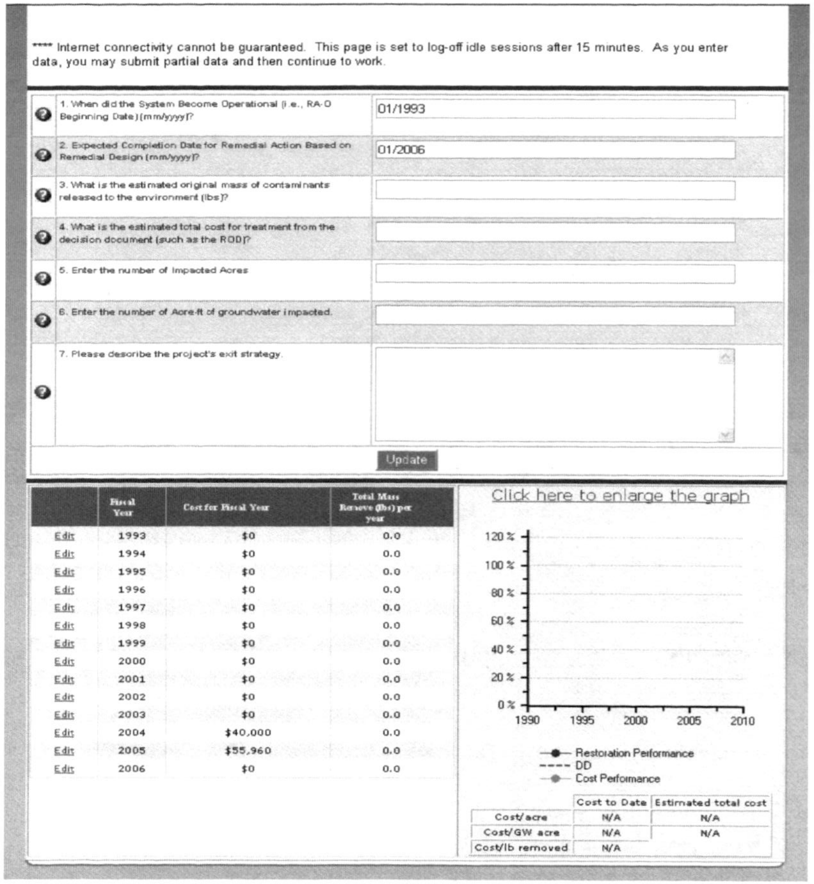

Integrity - Service - Excellence

U.S. AIR FORCE

Performance Tracking Tool - Chart

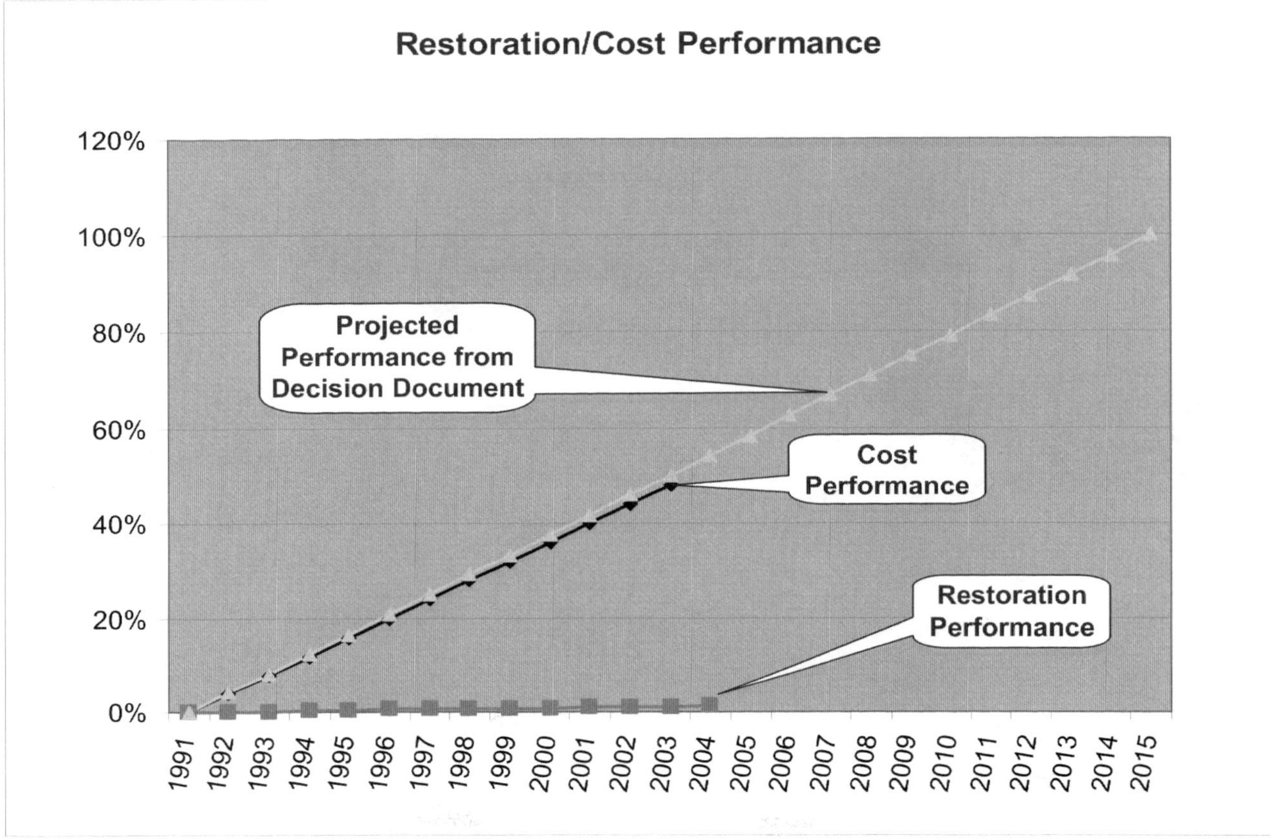

Integrity - Service - Excellence

U.S. AIR FORCE

RPO Benefits

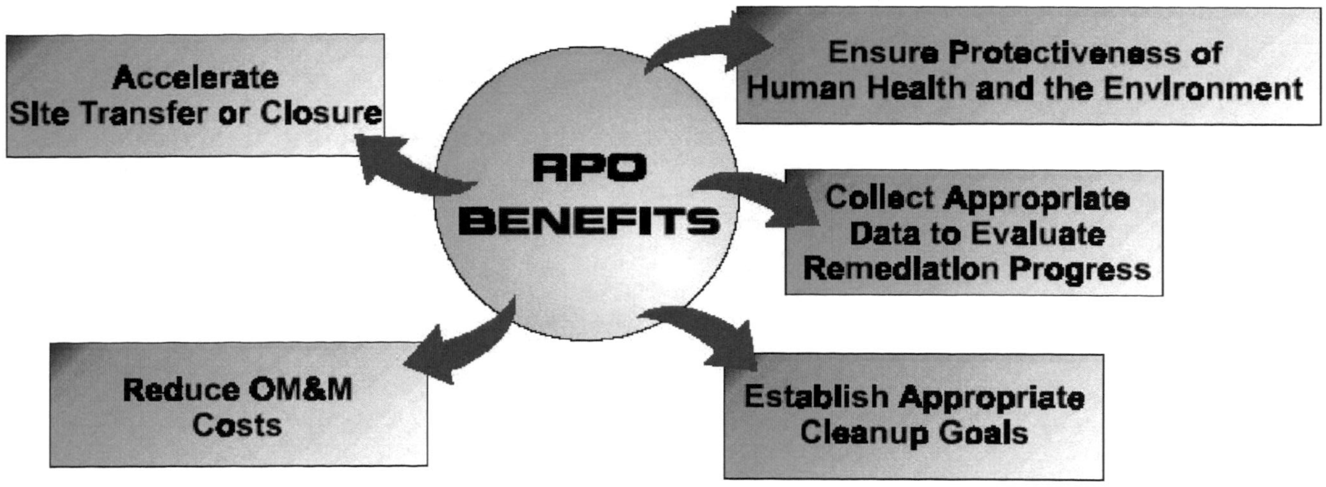

Integrity - Service - Excellence

U.S. AIR FORCE

The Result - RIPS

Results:

- Full Cost Visibility
- A Baseline
- The ability to track performance year to year

Room For Improvement

- RIPS could adequately track details of Record of Decision (ROD)
  - Basis for cleanup
    - Risk-based?
    - ARARs?
    - Regulatory Guidance
  - Contaminant(s) driving cleanup?
  - Receptors?
  - Matrix?
- RIPS could track the hydrogeologic factors
  - Site hydrogeologic description?
  - Hydrogeologic classification?
  - Complexity?
- RIPS could track Constraints to Remedial Effectiveness

Integrity - Service - Excellence

U.S. AIR FORCE

AFCEE RPO Outreach Office

- The RPO Outreach Office is the AF Service Center to:
  - Facilitate the RPO process
  - Act as repository to organize RPO information to include the RIPS database, RPO reports, and success stories/case studies
  - Develop tools to provide essential RPO information and services
  - Assist MAJCOMs and installations with requested contract and other support to accomplish RPO efforts
- AFCEE RPO Outreach Office Contact Info
  - http://www.hqafcee.brooks.af.mil/products/rpo/rpooutreach/rpo.asp
  - DSN 240-5294/4417
  - Commercial 210-536-5294/4417

U.S. AIR FORCE

Questions or Comments?

Integrity - Service - Excellence

U.S. AIR FORCE

RPO References

- AFCEE Guidance Jun 2001
  RPO Handbook
- ITRC Guidance Sep 2004
  RPO Identifying Opportunities for Enhanced and More Efficient Site Remediation

U.S. AIR FORCE

Session Acronyms

- AF – Air Force
- AFCEE – Air Force Center for Environmental Excellence
- AFI – Air Force Instruction
- CMS – Corrective Measures Study
- CY – Calendar Year
- DERP – Defense Environmental Restoration Program
- FS – Feasibility Study
- GTS – Geostatistical Temporal and Spatial
- ITRC – Interstate Technology & Regulatory Council
- LTM – Long-Term Monitoring
- MAJCOM – Major Air Force Commands
- MAROS – Monitoring and Remediation Optimization Software

Session Acronyms

- O&M – Operation and Maintenance
- PBC – Performance-Based Contracting
- PBM – Performance-Based Management
- RSC – Rapid Site Characterization
- RA – Remedial Action
- RIPS – RPO Inventory & Performance Software
- RPM – Restoration Program Managers
- RPO – Remedial Process Optimization
- SAF/IEE – Deputy Assistant Secretary of the Air Force for Environment, Safety and Occupational Health
- SC – Site Cleanup

254a Staged Electronic Data Deliverable Short Course

Anand Mudambi, Analytical Services Branch, US Environmental Protection Agency, 1200 Pennsylvania Ave NW, Mail Code 5102 G, Washington, DC 20460
The objective of the short course is to familiarize participants with the Staged Electronic Data Deliverable (SEDD). SEDD is now required by certain U.S. Environmental Protection Agency (EPA) and U.S. Army Corps of Engineers (USACE) contracts for the electronic reporting and management of environmental analytical data.

The short course is directed to people who are actively involved in the creation, use and implementation of SEDD. These would include personnel from laboratories and vendors who generate electronic deliverables for their clients as well as chemists from government agencies and contractors responsible for the review of data received from laboratories for environmental projects.

Participants are encouraged to bring laptops so that they can view and edit SEDD files as well use some SEDD applications. SEDD files and some applications will be provided.

SEDD is a non proprietary format for the electronic delivery of environmental analytical data and uses the Extensible Markup Language (XML). SEDD is designed to be agency and program neutral. SEDD incorporates a Data Element Dictionary along with a Valid Value list. Under SEDD, an electronic data deliverable from the laboratory is transmitted as an XML document.

Two software applications, the SEDD Parser Tool and the Automated Data Review (ADR) were developed under contract with the USACE Sacramento District in order to capture and review data in a timely and cost efficient manner. The SEDD Parser Tool captures data present in a SEDD file and allows migration of the data to ADR. The ADR software verifies compliance with project-specific data integrity and content rules and performs automated review of the data.

Data movement and storage issues will be discussed, including recent developments allowing transfer of reviewed data and ADR project library information in XML format for portability and long term storage.

RAPIC SITE CHARACTERIZATION (RSC)

EXTENDED ABSTRACT

Rapid Site Characterization (RSC) is a component of Performance Based Management (PBM) which is the Air Force's philosophy and approach in managing and assessing the performance of its Environmental Restoration Program. Accelerated characterization principles were originally developed by ASTM and the Interstate Technology Regulatory Council (ITRC). Based on the ASTM and ITRC guidance, RSC has been developed by the Air Force Center for Environmental Excellence (AFCEE) to emphasize the need to incorporate evolving scientific and technical information on site characterization actions from which remediation decisions can be made, and when conducted as part of Performance-Based Management principles protectiveness can be documented and completed in a reasonable time frame in a cost effective manner. The RSC's purpose is to provide environmental decision makers and practitioners with a tool to rapidly, cost-effectively, and reliably deliver a mental picture of current environmental conditions of a site. This mental picture is conveyed through a vehicle called a representative Conceptual Site Model (CSM). The CSM is the vehicle used to succinctly and clearly identify and depict the environmental problem at a site. RSC has six interlinking components (initial CSM, Dynamic Work plans, Real Time Data Analysis, Dynamic Document Review, Decision Logic, and Systematic Planning) with the central component being systematic planning. It is through systematic planning that the other components bring together the site understanding through the representative conceptual site model. This information becomes the basis for determining the necessity of site remediation. It is during the RSC that critical parameters for the design and optimization of the remedial system are identified and measured. The RSC uses a multi-discipline team to execute the various components. This core technical team includes analytical chemists, engineers, geologists, geochemists, geophysicists, hydrogeologists, risk assessors, and regulators. This core team helps with data management, QA/QC, risk assessment, fate and transport modeling, remedial action, community relations, and health and safety. The technical team will be responsible for: 1) Gathering all available information for the site, 2) developing an initial "conceptual" model for the site, 3) identifying the technical objectives and goals to be accomplished, 4) supervising the field effort, making adjustments to the Conceptual Site Model based on the data produced in the field, and 5) evaluating the conceptual model and decisions made with respect to federal, state, and local regulations. MAJCOMs and installations may request site-specific support for accelerated or Rapid Site Characterization through the AFCEE Project Manager and/or the appropriate directorate representative letting the contracts for the project or site. General information or questions can be answered by sending an email to the AFCEE Technology Transfer Outreach Office (TTO) at afcee.td.awag@brooks.af.mil. The phone number is commercial 210-536-4366 (DSN 240-4366).

HEADQUARTERS U.S. AIR FORCE

The Air Force Center for Environmental Excellence

AFCEE's Rapid Site Characterization Program

Ed Brown
AFCEE/TDE

Integrity - Service - Excellence

As of: 23 Sept 05

1

Outline

- Definition
  - PBM
  - RSC
- PBM Components
- RSC Components
- Purpose
- Process

Integrity - Service - Excellence

PBM Definition

Performance-Based Management

- A holistic and systematic results-oriented uncertainty, contingency, and information management assessment to expedite closure through the incorporation of:
  - Clear and accurate definition of the environmental problem through updated conceptual site model (CSM)
  - Established Exit Strategy
  - Development/Implementation of Decision Analyses & Logic
  - Identification & Tracking of performance-based Metrics
  - Performance-based Contracts

Interstate Technology Regulatory Council

PBM Definition

- **Performance-based Management is a strategic goal-oriented uncertainty management methodology that is implemented through systemic planning and dynamic decision-logic focused on the desired end-results**

PBM Components
A Defined Problem, Land Use, & Objectives
Process Optimization
Representative CSM
Contracting or Privatization Strategy
Expert Team & Communication
Established Land Use & Risk Mgmt. Strategy
Restoration Exit Strategy
Established ARAR Analysis Strategy
Restoration Decision Logic
ARAR Applicable or relevant & appropriate requirements
CSM Conceptual Site Model
As of: 23 Sept 05
Integrity - Service - Excellence
5

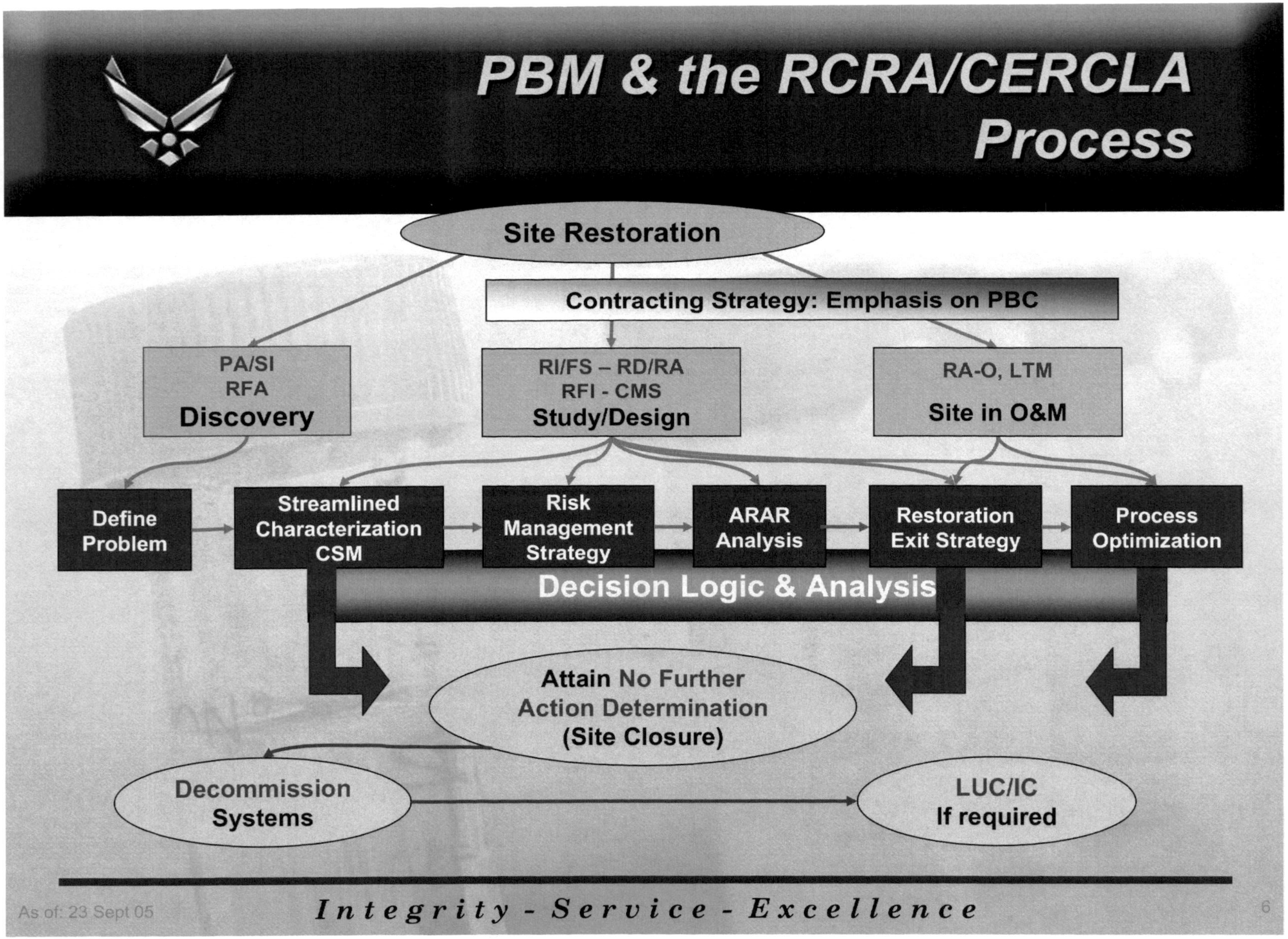
PBM & the RCRA/CERCLA Process
Site Restoration
Contracting Strategy: Emphasis on PBC
PA/SI
RFA
Discovery
RI/FS – RD/RA
RFI - CMS
Study/Design
RA-O, LTM
Site in O&M
Define Problem
Streamlined Characterization CSM
Risk Management Strategy
ARAR Analysis
Restoration Exit Strategy
Process Optimization
Decision Logic & Analysis
Attain No Further Action Determination (Site Closure)
Decommission Systems
LUC/IC
If required
As of: 23 Sept 05
Integrity - Service - Excellence
6

What is Rapid Site Characterization (RSC)?

- Accelerated procedure based on systematic decision logic to delineate contamination at complex sites and construct representative hydrogeologic conceptual site models

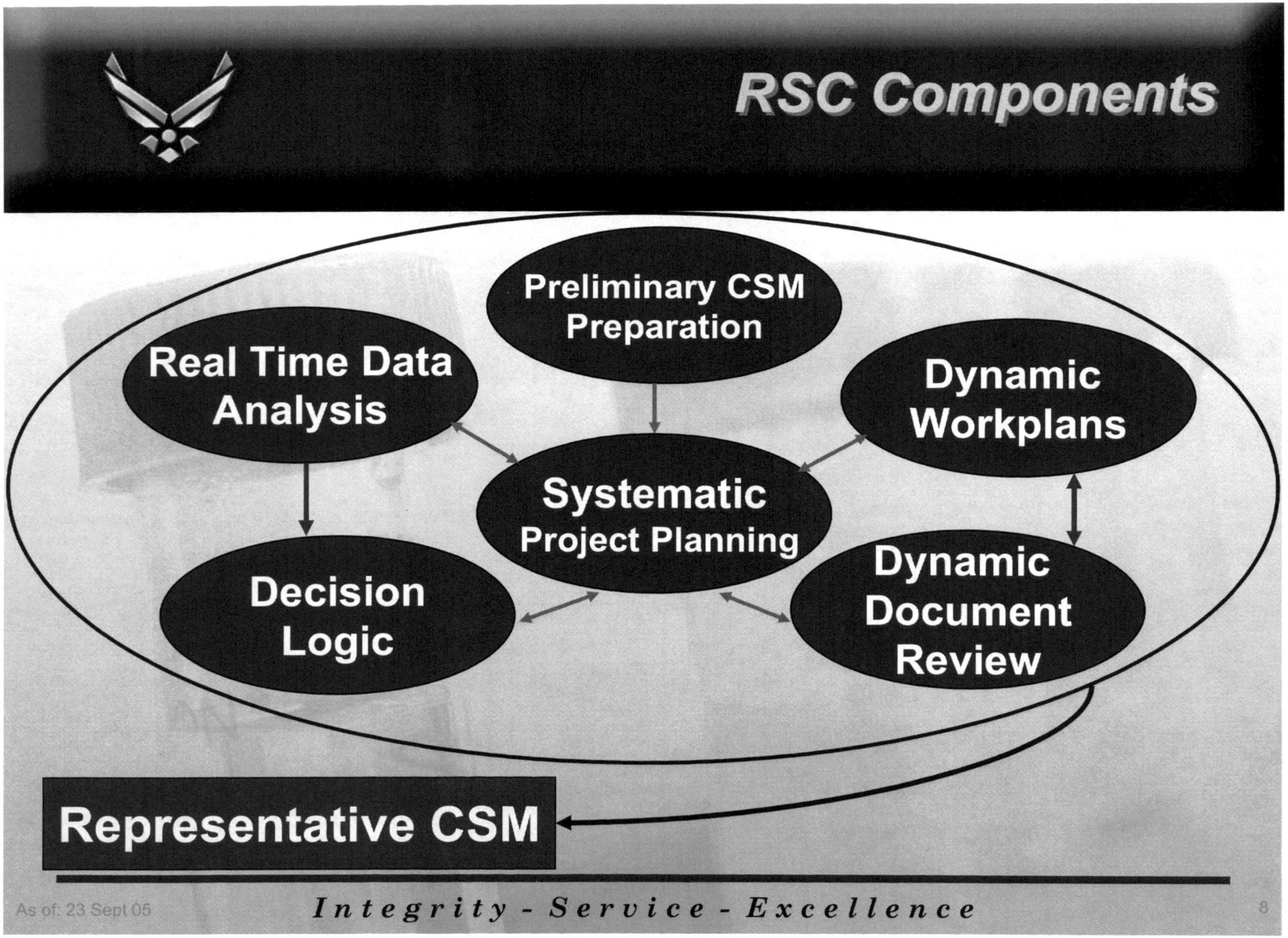
RSC Components
Preliminary CSM Preparation
Real Time Data Analysis
Dynamic Workplans
Systematic Project Planning
Decision Logic
Dynamic Document Review
Representative CSM
As of: 23 Sept 05
Integrity - Service - Excellence
8

Why use RSC?

- Create a functional restoration team
  - Face-to-face systematic planning more intensive, more far-reaching, than most imagine
- Empower field managers to make decisions
- RSC CSM captures heterogeneity & physical reality
- Analytical quality is NOT THE SAME AS Data quality
- A CSM depends on "Data representativeness"

- "Basic premise" plan before initiating field work
  - Every sample taken must lead to making a decision
  - Never take a sample that has no decision impact

Analytical Data Quality vs. Representativeness

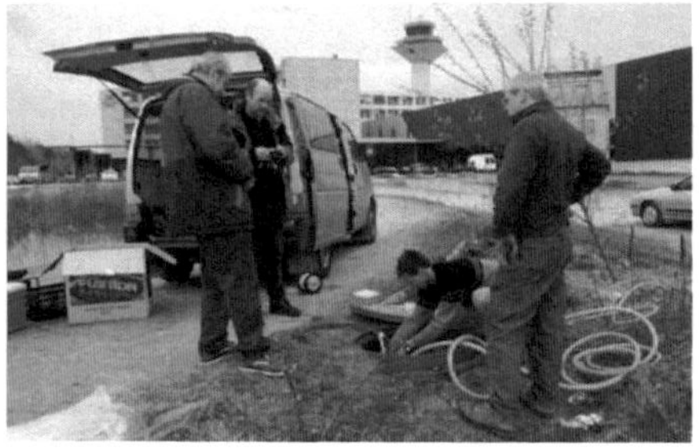

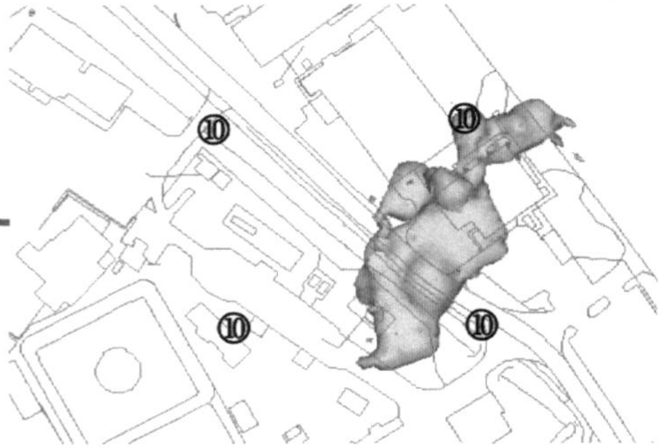

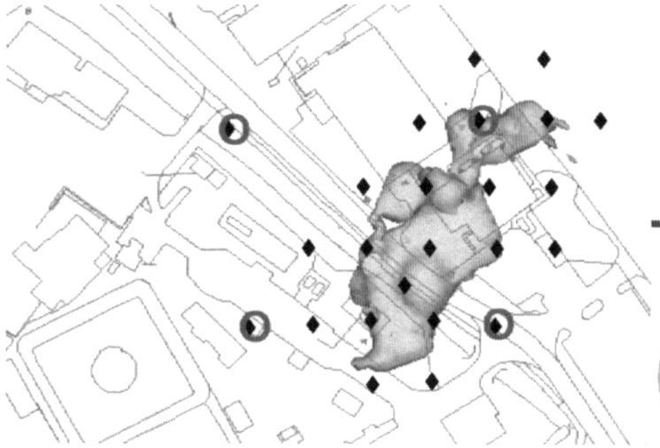

Four definitive quality data points generate a non-representative picture of the site

Screening quality data & sampling decision logic generates representative picture of the site

Goal: Representative Conceptual site Model

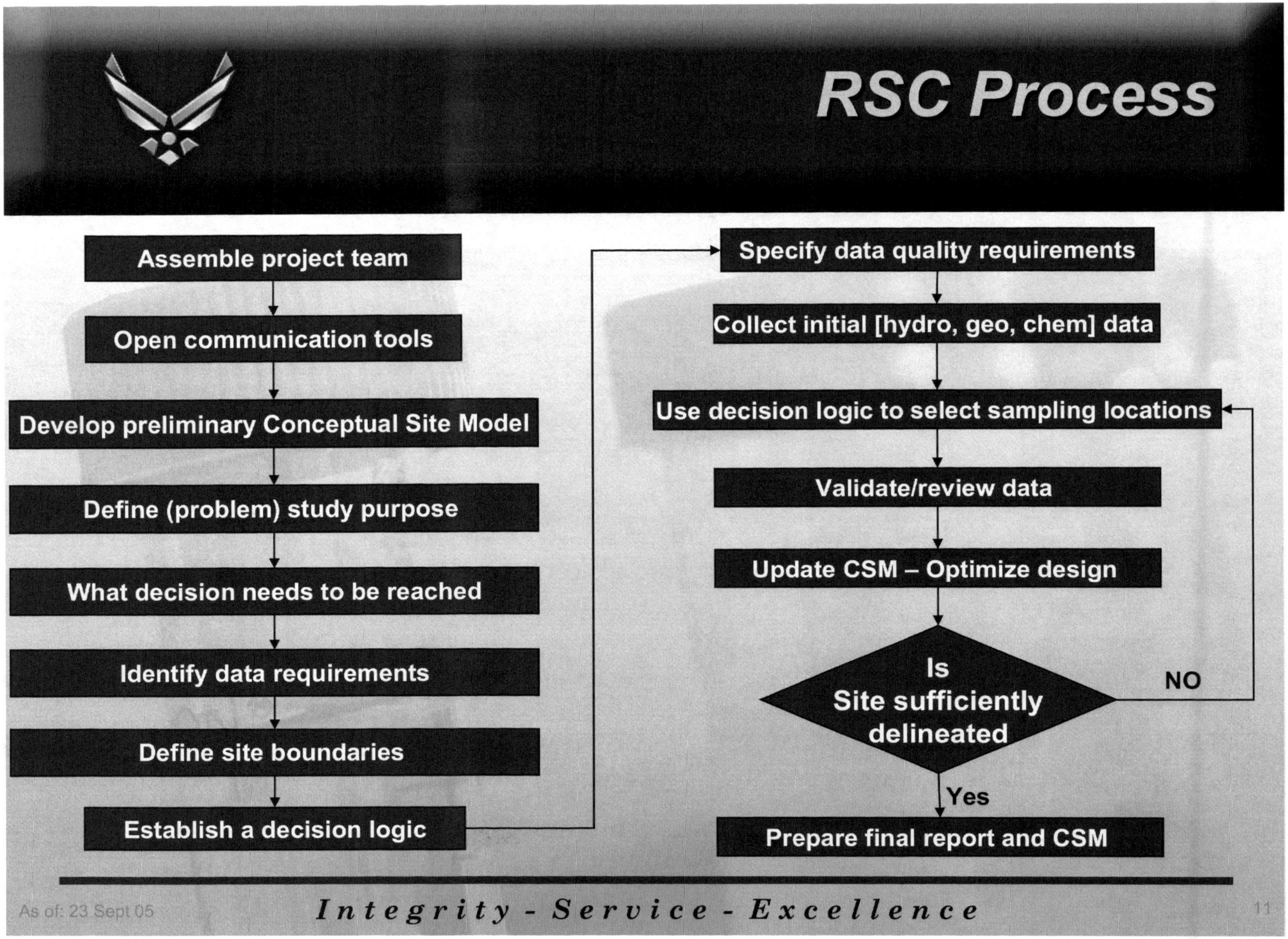
RSC Process
Assemble project team
Open communication tools
Develop preliminary Conceptual Site Model
Define (problem) study purpose
What decision needs to be reached
Identify data requirements
Define site boundaries
Establish a decision logic
Specify data quality requirements
Collect initial [hydro, geo, chem] data
Use decision logic to select sampling locations
Validate/review data
Update CSM – Optimize design
Is Site sufficiently delineated
NO
Yes
Prepare final report and CSM
As of: 23 Sept 05
Integrity - Service - Excellence
11

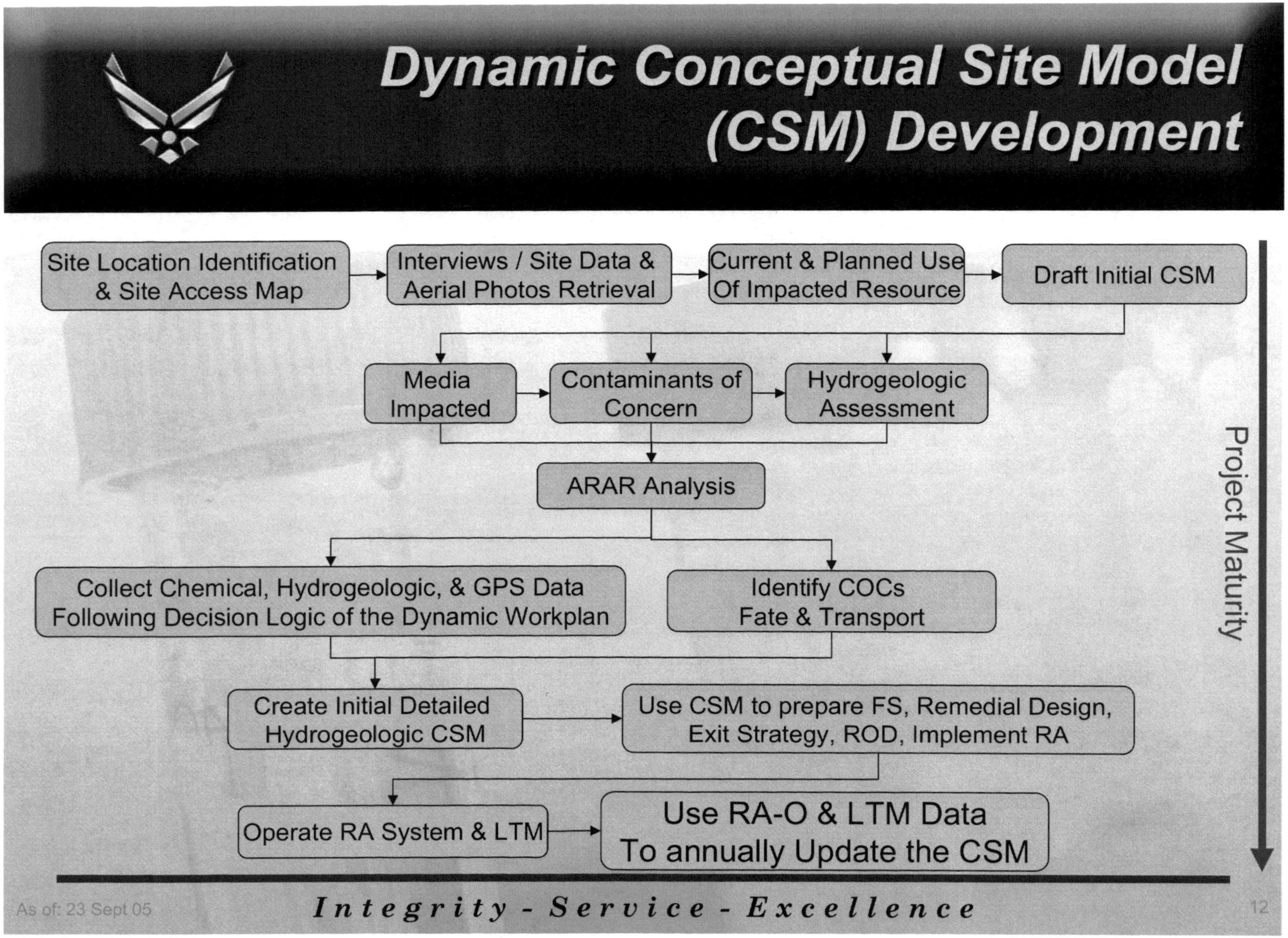
Dynamic Conceptual Site Model (CSM) Development
Site Location Identification & Site Access Map
Interviews / Site Data & Aerial Photos Retrieval
Current & Planned Use Of Impacted Resource
Draft Initial CSM
Media Impacted
Contaminants of Concern
Hydrogeologic Assessment
ARAR Analysis
Collect Chemical, Hydrogeologic, & GPS Data Following Decision Logic of the Dynamic Workplan
Identify COCs Fate & Transport
Create Initial Detailed Hydrogeologic CSM
Use CSM to prepare FS, Remedial Design, Exit Strategy, ROD, Implement RA
Operate RA System & LTM
Use RA-O & LTM Data To annually Update the CSM
Project Maturity
As of: 23 Sept 05
Integrity - Service - Excellence
12

References

- ASTM Active Standards D5730-04, E1689-95, E1912-98, and PS 3 – 95
- Regulatory Cooperation Work Group (ITRC) / American Society For Testing And Materials (ASTM) Partnership For Accelerated Characterization FY-97 Summary Report –Final December 1997

Want to Find Out More?

- Standard Guide for Accelerated Site Characterization for Confirmed or Suspected Petroleum Releases : http://www.astm.org/cgi-bin/SoftCart.exe/DATABASE.CART/REDLINE_PAGES/E1912.htm?E+mystore
- Interstate Technology & Regulatory Council TechReg Guideline for Triad: http://www.triadcentral.org/ref/ref/documents/SCM-1.pdf
- Regulatory Cooperation Work Group (ITRC) / American Society For Testing And Materials (ASTM) Partnership For Accelerated Characterization: http://www.itrcweb.org/Documents/ASC-1.pdf
- Performance Based Management and Rapid Site Characterization Guidance:

  http://www.afcee.brooks.af.mil/products/techtrans/

Questions?

Integrity - Service - Excellence

ACRONYMS AND ABBREVIATIONS

§	Section of regulation or statute
ACL	Alternate concentration limit
AFBCA	Air Force Base Conversion Agency
AFCEE	Air Force Center for Environmental Excellence
AFI	Air Force Instruction
AFRPA	Air Force Real Property Agency
AR	Administrative record
ARAR	Applicable or relevant and appropriate requirements
ASTM	American Society for Testing and Materials
BACT	Best available control technology
BCP	Base Closure Plan
BRAC	Base realignment and closure
BTEX	benzene, toluene, ethylbenzene and xylene
CERCLA	Comprehensive Environmental Response, Compensation, and Liability Act
CERFA	Community Environmental Response Facilitation Act
CFR	conceptual site model
CSM	Conceptual site model
CSSRA	chemical and site-specific risk assessment
CTC	Cost to complete
DD	Decision document
DERP	Defense Environmental Restoration Program
DNAPL	dense nonaqueous-phase liquid
DoD	Department of Defense
DQO	Data quality objective
EPA	Environmental Protection Agency
ERA	Environmental Restoration Account
ERP	Environmental Restoration Program
ERPIMS	Environmental Restoration Program Information Management
ESOH	Environment, Safety, and Occupational Health
FAR	Federal Acquisition Regulations
FOST	finding of suitability to transfer
FR	Federal Register
FS	Feasibility Study
IC	Institutional control
IRP	Installation Restoration Program
ITRC	Interstate Technology Regulatory Council
kg	kilogram
LNAPL	light nonaqueous-phase liquid
LTM	long-term monitoring
LTO	Long-term operations
LUC/IC	Land use control/institutional control
MAJCOM	Major Command

MAP	Management Action Plan
MCL	maximum contaminant level
MCLG	Maximum contaminant level goal
mg	milligram
MNA	Monitored natural attenuation
MOU	Memorandum of Understanding
NAPL	non-aqueous-phase liquid
NCP	National Contingency Plan
NESHAP	National Emission Standards for Hazardous Air Pollutants
NICRM	Natural Infrastructure Capability and Requirements Management
NPDES	National Pollution Discharge Elimination System
O&M	Operations and maintenance
OA	Objectives assessment
OM&M	operation maintenance and monitoring
OPS	operating property and successfully
OSWER	Office of Solid Waste and Emergency Response
P&PR	Performance and Protectiveness Review
PA	Preliminary assessment
PBC	Performance-based contract
PBM	Performance-based management
PBMS	performance based measurement system
PBRI	Performance-base remedial investigation
PC	personal computer
PFD	process flow diagram
POTW	publicly owned treatment works
ppb	part per billion
ppmv	part-per-million-by-volume
PRB	Permeable reactive barrier
PRP	Potentially responsible party
PTT	Performance Tracking Tool
QA	Quality assurance
QAPP	Quality Assurance Project Plan
QC	Quality control
RAB	restoration advisory board
RAO	Remedial action objective
RA-O	Remedial action operations
RBCA	risk-based corrective action
RBSL	risk-based screening level
RC	Resource Capability (Model)
RCRA	Resource Conservation and Recovery Act
RF	Radio frequency
RfR	Ready for reuse
RI	remedial investigation
RI/FS	remedial investigation/feasibility study
RIPS	RPO Inventory and Performance Software

ROD	Record of Decision
RPM	Remedial project manager
RPO	remedial process optimization
RSE	Remedial Systems Evaluation
RSO	remedial system optimization
RV	Resource Valuation (Model)
SAF	Secretary of the Air Force
SARA	Superfund Amendments and Reauthorization Act of 1986
SEDD	Staged Electronic Data Deliverables
SI	Site Investigation
SOO	Statement of objectives
SOW	Statement of work
SSI	Streamlined site investigation
SSL	soil screening level
SSTL	site-specific target level
SVE	Soil vapor extraction
System	
TCE	trichloroethylene
TI	technical impracticability
TIO	Technology Innovation Office (now OSWTI)
TPH	total petroleum hydrocarbon
UFP	Uniform federal policy
USACE	US Army Corps of Engineers
USAF	US Air Force
USCA	United States Code Anotated
USEPA	U.S. Environmental Protection Agency
UST	underground storage tank
VMP	vapor monitoring point
VOC	volatile organic compound

American Institute of Chemical Engineers
3 Park Avenue
New York, New York 10016-5991

ISBN: 978-1-60423-531-9